The table of logarithms of trigonometric functions (Table II), the most important one for trigonometry, has a number of new features. The proportional parts are tabulated for each second from 0″ to 60″, and bold-faced numbers have been so used as to avoid ambiguity. Whenever there is a choice of two numbers one of which is written in bold face, the bold-faced number is always chosen. The simplicity of operation introduced by this plan gives a gain both in speed and in accuracy. In the table proper all six functions are tabulated, and boldfaced numbers are used in such a way as to enable the user to locate approximate position by using them only.

Log and Trig Tables

Lyman M. Kells, PhD.
Willis F. Kern
James R. Bland,
Professor of Mathematics
United States Naval Academy

McGraw-Hill Book Company, Inc.
New York Toronto London

PRINTED IN THE UNITED STATES OF AMERICA

14151617 MUMU 98

ISBN 07-033601-6

PREFACE

A table of logarithms should be accurate, it should be easy to understand, and it should be as easy to use as possible. The authors, in the tables offered here, have attempted to make improvements along these three lines.

The tables used in trigonometry and its applications have been checked many times and have been carefully read against other tables. If, in spite of this thoroughness in compilation, errors are discovered, the authors would appreciate having them pointed out.

Frequently students fail to understand the process of linear interpolation. It is explained in this book by means of a simple diagram which gives the idea almost at a glance.

The table of logarithms of trigonometric functions (Table II), the most important one for trigonometry, has a number of new features. The proportional parts are tabulated for each second from 0″ to 60″, and bold-faced numbers have been so used as to avoid ambiguity. Whenever there is a choice of two numbers one of which is written in bold face, the bold-faced number is always chosen. The simplicity of operation introduced by this plan gives a gain both in speed and in accuracy. In the table proper all six functions are tabulated, and bold-faced numbers are used in such a way as to enable the user to locate approximate position by using them only. It is believed that the gains due to these innovations are decidedly worth while.

LYMAN M. KELLS.
WILLIS F. KERN.
JAMES R. BLAND.

ANNAPOLIS, MD.,

CONTENTS

TABLE I

COMMON LOGARITHMS OF NUMBERS

1. Introduction.* *The power L to which a given number b must be raised to produce a number N is called the logarithm of N to the base b.* This relation expressed in symbols is

$$b^L = N.$$

It appears at once that b must not be unity and it must not be negative. In the following set of tables, 10 is used as base. This system is called the *common system* or the *Briggs system.* Another important system, called the *natural system,* has e as base, where $e = 2.71828$ accurate to six figures.

2. Characteristic and mantissa. The common logarithm of any real, positive number may be written as an integer, positive or negative, plus a positive decimal fraction. The integral part is called the *characteristic* and the decimal part the *mantissa.* The characteristic may be written by using the following rules:

Rule 1. *The characteristic of the common logarithm of a number greater than* 1 *is obtained by subtracting* 1 *from the number of digits to the left of the decimal point.*

Rule 2. *The characteristic of the common logarithm of a positive number less than* 1 *is negative and its magnitude is obtained by adding* 1 *to the number of zeros immediately following the decimal point.*

If the characteristic of a number is $-n$ (n positive), it should be written in the form $(10 - n) - 10$. *To obtain directly the logarithm of a number less than* 1, *subtract from* 9 *the number of zeros immediately following the decimal point, and write the result before the mantissa and* -10 *after it.*

The method of finding the mantissa of the logarithm of a number will be explained in the succeeding articles.

* Since the theory of logarithms is treated completely in algebra and in trigonometry, only the actual manipulation of the tables is explained here.

EXERCISES

Verify the characteristic of the logarithm of each of the numbers N written below

	N	$\log N$		N	$\log N$
1.	6.830	0.83442.	**8.**	58.73	1.76886.
2.	68.30	1.83442.	**9.**	0.6740	9.82866 − 10.
3.	6830	3.83442.	**10.**	0.007500	7.87506 − 10.
4.	683,000	5.83442.	**11.**	6.870×10^5	5.83696.
5.	0.7860	9.89542 − 10.	**12.**	5.860×10^{-4}	6.76790 − 10.
6.	0.007860	7.89542 − 10.	**13.**	3.990×10^{-6}	4.60097 − 10.
7.	0.0007860	6.89542 − 10.	**14.**	7.330×10^2	2.86510.

3. To find the mantissa. Special case. The mantissa, or decimal part of the logarithm of a number, depends only on the sequence of the digits and not on the position of the decimal point. Table I lists the mantissas, accurate to five decimal places, of the logarithms of all integers from 1 to 10,000.

The change in the mantissas of the logarithms is so slow that the first two figures do not change for several lines of the table. Consequently the appropriate first two figures are printed in the first column before the first full row to which they apply. Also the appropriate first two figures appear at the left of the first line of mantissas on each page. An asterisk in any row indicates that the first two figures are to be found at the left of the next row.

To find the mantissa of the logarithm of a number locate the first three digits of this number in the left-hand column headed N and the fourth digit in the row at the top of the page. Then the mantissa of the given number containing four significant figures is in the row whose first three figures are the first three significant figures of the given number, and in the column headed by the fourth. Thus to find the logarithm of 76.64 find 766 in the column headed N, follow the corresponding row to the entry in the column headed by 4. This entry 88446 represents the mantissa required. Hence we have

$$\log 76.64 = 1.88446. \quad \textit{Ans.}$$

EXERCISES

Verify the logarithms in the exercise of §2.

4. Interpolation. When a number contains a fifth significant figure, we find the logarithm corresponding to the first four figures as in §3 and then add an increment obtained by a process called interpolation. This process is based on the assumption that *for relatively small changes in the number N the changes in log N are proportional to the changes in N.* The following example will serve to illustrate the process of interpolation.

The expression *tabular difference* will be used frequently in what follows. The tabular difference, when used in connection with a table, means the result of subtracting the lesser of two successive entries from the greater.

Example. Find log 235.47.

Solution. We first find the logarithms in the following form and then compute the difference indicated:

$$\left.\begin{array}{l} \left.\begin{array}{l}\log 235.40\\ \log 235.47\end{array}\right\}7 \\ \log 235.50 \end{array}\right\}10 \quad \left.\begin{array}{l} \left.\begin{array}{l}= 2.37181\\ = \ ?\end{array}\right\}d \\ = 2.37199 \end{array}\right\}18 \text{ (tabular difference*)}$$

By the principle of proportional parts, we have

$$\frac{7}{10} = \frac{d}{18}, \qquad \text{or} \qquad d = \frac{7}{10}(18) = 12.6 = 13 \text{ (nearly)}.$$

Adding 0.00013 to 2.37181, we obtain

$$\log 235.47 = 2.37194. \quad \textit{Ans.}$$

The increment 12.6 was rounded off to 13 because we are not justified in writing more than five decimal places in the mantissa.

The essence of this procedure is embodied in the following statement. To find the logarithm of a number composed of five significant figures, first find the logarithm corresponding to the first four figures and to it add one-tenth of the tabular difference multiplied by the fifth digit.

To shorten the process of interpolation, 10^5 times each tabular difference occurring in the table has been multiplied by 0.1, 0.2, . . . 0.9, and the results have been tabulated on the right-hand sides of the pages on which these differences occur. The abbreviation Prop. Parts written at the top of the page over these small tables abbreviates the words *proportional parts*. To interpolate in the example just solved, locate the Prop. Parts table headed 18 and find opposite 7 in its left-hand column the entry 12.6 (=13 nearly). In general, this difference should not be computed but should be obtained from the number opposite the fifth digit in the appropriate table of proportional parts.

EXERCISES

Verify the following logarithms:

1. log 7012.6 = 3.84588.
2. log 54.725 = 1.73819.
3. log 0.87364 = 9.94133 − 10.
4. log 3.7245 = 0.57107.
5. log 0.00065931 = 6.81909.
6. log 25.819 = 1.41194.
7. log 2.3454 = 0.37022.
8. log 0.056321 = 8.75067 − 10.
9. log 4,574,000 = 6.66030.
10. log 568.91 = 2.75504.
11. log 4.3965×10^5 = 5.64311.
12. log 10.905 = 1.03763.
13. log 0.0025725 = 7.41036.
14. log 0.000032026 = 5.50550 − 10.

5. To find the number corresponding to a given logarithm. If log $N = L$, the number N is called the *antilogarithm* of L. The sequence of

* For convenience the decimal point has been omitted.

digits of a number N corresponding to a given logarithm L is found from its mantissa, and the decimal point is then placed in accordance with the rules of §2.

Example. Given $\log N = 1.60334$, find N.

Solution. The mantissa .60334 lies between the entries .60325 and .60336 of Table I. Using the table and computing the differences indicated, we write the following form:

$$\left.\begin{matrix}\left.\begin{matrix}1.60325\\1.60334\end{matrix}\right\}9\\1.60336\end{matrix}\right\}11 \begin{matrix}= \log 40.110\\= \log N\\= \log 40.120\end{matrix}\ \left.\begin{matrix}\left.\begin{matrix}\\ \\ \end{matrix}\right\}x\\ \\ \end{matrix}\right\}10$$

Assuming that changes in the logarithm are proportional to the corresponding changes in the number, we write

$$\frac{9}{11} = \frac{x}{10}, \qquad \text{or} \qquad x = 10\left(\frac{9}{11}\right) = 8 \text{ (nearly)}.$$

Hence

$$N = 40.118. \quad \textit{Ans.}$$

The essence of the process of interpolation is indicated in the foregoing procedure. However, in practice, the student should always interpolate by using the table of proportional parts. The fifth figure 8 should have been obtained from the table of proportional parts. In the small Prop. Parts table corresponding to the tabular difference 11, we read the fifth figure 8 in the left-hand column opposite the entry 8.8, the entry nearest to 9.

EXERCISES

Verify the following antilogarithms:

1. 3.57351 = log 3745.5.
2. 2.82315 = log 665.50.
3. 0.12112 = log 1.3217.
4. 1.92594 = log 84.321.
5. 9.47954 − 10 = log 0.30167.
6. 8.65636 − 10 = log 0.045327.
7. 0.37976 = log 2.3975.
8. 4.76224 = log 57842.
9. 6.51738 − 10 = log 0.00032914.
10. 1.49715 = log 31.416.
11. 4.21691 − 10 = log 16478.
12. 5.09873 = log 125520.
13. 9.27951 − 10 = log 0.19033.
14. 7.88000 − 10 = log 0.0075858.

TABLE II

LOGARITHMS OF TRIGONOMETRIC FUNCTIONS

6. Table of logarithms of trigonometric functions. Table II gives the logarithms of the sines, cosines, tangents, cotangents, secants, and cosecants of angles at intervals of 1′ from 0° to 90°. The names of the functions written at the top of any page apply to angles having the number of degrees written at the top of the page, and the function names written at the bottom apply to angles having the number of degrees written at the bottom. The left-hand or the right-hand minute column applies according as the number of degrees in the angle is written on the left side or on the right side of the block of numbers under consideration.

For example, to find log sin 32° 46′, we find the page at the top of which 32° appears, find the row containing 46 in the left-hand minute column, and read 73337 in this row and in the column headed *l* sin. Hence log sin 32° 46′ = 9.73337 − 10. The number 9 was found at the head of the *l* sin column and the number −10 is to be applied to every logarithm in the table. Again, to find log tan 142° 36′, find the page at the top of which 142° appears, find the row containing 36 in the right-hand minute column, and read 88341 in this row and in the column headed *l* tan. Hence log tan 142° 36′ = (−) 9.88341 − 10. The minus sign in parentheses before the log indicates that a negative number is under consideration. The characteristic was obtained as in the first example.

EXERCISES

Verify the following:

1. log sin 37° 27′ = 9.78395 − 10.
2. log tan 36° 41′ = 9.87211 − 10.
3. log cot 28° 16′ = 0.26946.
4. log cos 62° 20′ = 9.66682 − 10.
5. log csc 69° 54′ = 0.02729.
6. log sin 131° 10′ = 9.87668 − 10.
7. log tan 142° 27′ = (−) 9.88577 − 10.
8. log sec 134° 47′ = (−) 0.15216.
9. log cos 45° 47′ = 9.84347 − 10.
10. log csc 135° 13′ = (−) 0.15216.
11. log cot 132° 0′ = (−) 9.95444 − 10.

7. Given the angle, to find the logarithm of a trigonometric function. The principles involved here are the same as those involved in finding logarithms and antilogarithms of numbers. Interpolation for seconds is accomplished by direct interpolation or by using the columns headed d 1′ and the columns headed proportional parts. The following example will illustrate the procedure.

Example. Find log tan 65° 42′ 17″.

Solution. Using the table to find logarithms and computing differnces, we write the following form:

$$\left.\begin{array}{l}\left.\begin{array}{l}\log\tan 65^\circ\,42'\,00''\\ \log\tan 65^\circ\,42'\,17''\end{array}\right\}17''\\ \log\tan 65^\circ\,43'\,00''\end{array}\right\}60'' \qquad \left.\begin{array}{l}\left.\begin{array}{l}=0.34533\\ \quad ?\end{array}\right\}x\\ =0.34566\end{array}\right\}33$$

Hence assuming that, for small changes, change of logarithm is proportional to change of angle, we have

$$\frac{x}{33} = \frac{17}{60}, \qquad \text{or} \qquad x = 33\left(\frac{17}{60}\right) = 9.35 = 9 \text{ (nearly)}.$$

Therefore

$$\log\tan 65^\circ\,42'\,17'' = 0.34533 + 0.00009 = 0.34542. \quad \textit{Ans.}$$

The essence of the process of interpolation is indicated in the foregoing procedure. However, in practice, the student should always interpolate by using the columns headed d 1′ and the proportional parts column.

Each entry in the column headed d 1′ gives the difference of the logarithms between which it is spaced in each of the adjacent columns. In each column headed by *proportional parts* appears $\frac{1}{60}$, $\frac{2}{60}$, $\frac{3}{60}$ . . . of the number heading the column. Hence the difference 9 to be applied in the case of the foregoing example is found in the proportional parts column headed by 33 (the tabular difference for 1′ written between 0.34533 and 0.34566) and in the row with the 17 of the seconds column. Again, to find log cot 10° 28′ 36″, we find the entry 73345 for log cot 10° 28′, note the appropriate number 71 in the adjacent column headed d 1′, enter the proportional parts column headed by 71, read in this column 43 opposite the 36 of the seconds column; subtract 43 from 73345, and write log cot 10° 28′ 36″ = 0.73302.

It is worthy of note that *the changes of logarithms due to the seconds of an angle* must be added or subtracted according as the value of the function for angles near the one under consideration is increasing or decreasing with increasing angle.

EXERCISES

Verify the following:

1. log sin 35° 17′ 8″ = 9.76166 − 10.
2. log cos 48° 24′ 21″ = 9.82207 − 10.
3. log sec 142° 37′ 15″ = (−) 0.09984.

4. log csc 56° 21′ 57″ = 0.07956.
5. log cot 23° 16′ 50″ = 0.36626.
6. log csc 128° 47′ 52″ = 0.10826.
7. log tan − 69° 38′ 54″ = (−) 0.43070.
8. log sin 197° 36′ 57″ = 9.48092 − 10.
9. log sin 137° 45′ 22″ = 9.82756 − 10.
10. log cos 137° 45′ 22″ = (−) 9.86940 − 10.
11. log sin 209° 32′ 50″ = 9.69297 − 10.
12. log cos 330° 27′ 10″ = 9.93949 − 10.

8. Given the logarithm of a trigonometric function, to find the angle. The following example will indicate the procedure necessary to find the angle when the logarithm of a trigonometric function of the angle is given:

Example. Find θ if log cos θ is 9.85391 − 10.

Solution. Using the table to find logarithms and computing differences, we write the following form:

$$\left.\begin{array}{l}\left.\begin{array}{l}\log\cos 44^\circ\,24'\,00''\\ \log\cos 44^\circ\,24'\;\;?''\end{array}\right\}x\\ \log\cos 44^\circ\,25'\,00''\end{array}\right\}60'' \quad \begin{array}{l}= 9.85399\\ = 9.85391\\ = 9.85386\end{array} \quad \begin{array}{l}\}8\\ \end{array}\;\}13$$

Hence

$$\frac{x}{60} = \frac{8}{13}, \qquad \text{or} \qquad x = \frac{8}{13}(60) = 37'' \text{ (nearly)},$$

and

$$\theta = 44^\circ\,24'\,37''. \quad \textit{Ans.}$$

The essence of the process of interpolation is indicated in the foregoing procedure. In practice, however, the columns headed d 1′ and the proportional parts columns should be used in interpolation. Thus, to find θ in the example just considered, we first find 44° 24′ and difference 8 as above, then read 13 in the column headed d 1′ adjacent to and slightly below the entry 85399, enter the corresponding proportional parts column, opposite the bold-faced one of the five 8's tabulated read 37″ in the seconds column, and then write θ = 44° 24′ 37″.

When finding the number of seconds in an angle corresponding to a given logarithm of a trigonometric function, the student may find several identical entries in the proportional parts column involved. In this case, and *in any case where there is a choice between two or more entries one of which is printed in* **bold face,** *always give preference to the* **bold-faced** *entry.*

EXERCISES

Find the value of θ less than 360° in the following:

1. log sin θ = 9.96162 − 10. *Ans.* 66° 16′ 0″ and 113° 44′ 0″.
2. log cos θ = 9.99537 − 10. *Ans.* 8° 21′ 0″ and 351° 39′ 0″.
3. log cot θ = 0.52368. *Ans.* 16° 40′ 13″ and 196° 40′ 13″.

4. log tan θ = 9.50368 − 10. *Ans.* 17° 41′ 18″ and 197° 41′ 18″.
5. log cos θ = 9.96301 − 10. *Ans.* 23° 18′ 48″ and 336° 41′ 12″.
6. log sin θ = 9.84963 − 10. *Ans.* 45° 1′ 9″ and 134° 58′ 51″.
7. log cot θ = 9.50064 − 10. *Ans.* 72° 25′ 38″ and 252° 25′ 38″.
8. log tan θ = 0.96236. *Ans.* 83° 46′ 34″ and 263° 46′ 34″.
9. log sec θ = 0.12358. *Ans.* 41° 12′ 22″ and 318° 47′ 38″.
10. log csc θ = 0.71238. *Ans.* 11° 10′ 53″ and 168° 49′ 7″.

9. Angles near 0° and 90°. When angles are near 0° or near 90°, interpolation based on the assumption of proportional change in angle and logarithm may give results considerably in error. For this reason it is convenient to introduce the functions S and T defined by the equations $S = \alpha/\sin\alpha$ and $T = \alpha/\tan\alpha$. The relative change of the functions S and T with respect to α is very small when α is less than 3° and, as a consequence, the required accuracy of the results is obtained by using them. On the first three pages of Table II the columns headed log S* and log T give the common logarithms of S and T, respectively.

The following formulas apply when the angle involved is less than 3°:

1. For angles less in magnitude than 3°.

(*a*) $\log\sin\alpha = \log\alpha''$† $- \log S$.
(*b*) $\log\tan\alpha = \log\alpha'' - \log T$.
(*c*) $\log\cot\alpha = \text{colog}\,\alpha'' + \log T$,
$= \text{colog}\tan\alpha$.
(*d*) $\log\csc\alpha = \text{colog}\,\alpha'' + \log S$.
(*e*) $\log\alpha'' = \log\sin\alpha + \log S$.
(*f*) $\log\alpha'' = \log\tan\alpha + \log T$.
(*g*) $\log\alpha'' = \text{colog}\cot\alpha + \log T$.
(*h*) $\log\alpha'' = \text{colog}\csc\alpha + \log S$.

2. For angles α such that $90° - \alpha$‡ is less in magnitude than 3°.

(*i*) $\log\cos\alpha = \log(90° - \alpha)'' - \log S$.
(*j*) $\log\cot\alpha = \log(90° - \alpha)'' - \log T$.
(*k*) $\log\tan\alpha = \text{colog}\,(90° - \alpha)'' + \log T$,
$= \text{colog}\cot\alpha$.
(*l*) $\log\sec\alpha = \text{colog}\,(90° - \alpha)'' + \log S$.
(*m*) $\log(90° - \alpha)'' = \log\cos\alpha + \log S$.
(*n*) $\log(90° - \alpha)'' = \log\cot\alpha + \log T$.
(*o*) $\log(90° - \alpha)'' = \text{colog}\tan\alpha + \log T$.
(*p*) $\log(90° - \alpha)'' = \text{colog}\sec\alpha + \log S$.

To find θ when log sin θ = 8.46932 − 10, we first find in the column headed *l* sin the entry nearest to 8.46932, namely, 8.46799. On one side of 8.46799 we read log S = 5.31449, and on the other 1° 41′ = 6060″. Hence, using formula (*e*), we write log α = 8.46932 − 10 + 5.31449 =

* The function log S is often written cpl S, and the function log T, is written cpl T.

† The symbol log α'' means in this connection the logarithm of the number of seconds in the angle.

‡ Since $\cos\alpha = \sin(90° - \alpha)$, in this case $S = \dfrac{(90° - \alpha)''}{\sin(90° - \alpha)}$.

3.78381. Therefore $\alpha = 6078.7''$. Since $1° \ 41' = 6060''$, $6078.7'' = 1° \ 41' \ 19''$.

EXERCISES

Verify the following:

1. $\log \sin 0° \ 44' \ 13'' = 8.10930 - 10$.

2. $\log \cos 89° \ 21' \ 31'' = 8.04899 - 10$.

3. $\log \tan 0° \ 32' \ 23'' = 7.97406 - 10$.

4. $\log \cot 0° \ 25' \ 56'' = 2.12241$.

5. $\log \tan 1° \ 10' \ 9'' = 8.30981 - 10$.

6. $\log \cot 89° \ 3' \ 11'' = 8.21824 - 10$.

7. $\log \cos 88° \ 41' \ 20'' = 8.35948 - 10$.

8. $\log \sin 0° \ 59' \ 8'' = 8.23554 - 10$.

9. $\log \tan 1° \ 29' \ 10'' = 8.41403 - 10$.

10. $\log \sec 88° \ 16' \ 10'' = 1.52000$.

Verify the following:

11. $\log \cos \theta = 8.32967 - 10$; $\theta = 88° \ 46' \ 33''$ and $271° \ 13' \ 27''$.

12. $\log \tan \theta = 8.11584 - 10$; $\theta = 0° \ 44' \ 53''$ and $180° \ 44' \ 53''$.

13. $\log \sin \theta = 8.23468 - 10$; $\theta = 0° \ 59' \ 1''$ and $179° \ 0' \ 59''$.

TABLE III

NATURAL TRIGONOMETRIC FUNCTIONS

10. Table of natural values of trigonometric functions. Table III contains the numerical values of the sines, cosines, tangents, and cotangents of angles from 0° to 90° at intervals of 1′. In the case of an angle in the range from 0° to 45°, the number of degrees in the angle and the names of the functions are found at the top of the page and the left-hand minute column applies; in the case of angles in the range from 45° to 90°, the number of degrees in the angle and the names of the functions are found at the bottom of the page and the right-hand minute column applies. Interpolation must be carried out without the aid of difference columns or tables of proportional parts.

The following examples illustrate the method of using the tables.

Example 1. Find sin 68° 28′.

Solution. We first find the page at the bottom of which 68° appears and then find the row of the 68° block containing 28′ in the right-hand minute column. In this row and in the column having sin at its foot we find 020 to which we must prefix 0.93 to obtain sin 68° 28′ = 0.93020.

Example 2. Find sin 38° 38′ 27″.

Solution. Using the tables and computing differences, we find the values exhibited in the following form:

$$\left.\begin{array}{l}\left.\begin{array}{l}\sin 38^\circ\ 38'\ 00'' \\ \sin 38^\circ\ 38'\ 27''\end{array}\right\}27'' \\ \sin 38^\circ\ 39'\ 00''\end{array}\right\}60'' \quad \left.\begin{array}{l}\left.\begin{array}{l}= 0.62433 \\ = \ ?\end{array}\right\}x \\ = 0.62456\end{array}\right\}23$$

Hence

$$\frac{x}{23} = \frac{27}{60}, \qquad \text{or} \qquad x = \left(\frac{27}{60}\right)23 = 10 \text{ (nearly)}.$$

Therefore

$$\sin 38^\circ\ 38'\ 27'' = 0.62433 + 0.00010 = 0.62443. \quad \textit{Ans.}$$

Example 3. If cot θ = 0.37806, find θ.

Solution. Using the tables and computing differences, we find the values exhibited in the following form:

$$\left.\begin{array}{l}\left.\begin{array}{l}\cot 69^\circ\ 17'\ 00'' \\ \cot \quad ?\end{array}\right\}x \\ \cot 69^\circ\ 18'\ 00''\end{array}\right\}60 \quad \left.\begin{array}{l}\left.\begin{array}{l}= 0.37820 \\ = 0.37806\end{array}\right\}14 \\ = 0.37787\end{array}\right\}33$$

Hence

$$\frac{x}{60} = \frac{14}{33}, \quad \text{or} \quad x = \frac{14}{33}(60) = 25'' \text{ (nearly), and } \theta = 69° \ 17' \ 25''. \quad \textit{Ans.}$$

Since cot θ is positive in the third quadrant, we may also write an answer $180° + 69° \ 17' \ 25'' = 249° \ 17' \ 25''$. *Ans.*

EXERCISES

Verify the following:

1. $\sin 53° \ 42' \ 0'' = 0.80593$.

2. $\cos 31° \ 53' \ 9'' = 0.84911$.

3. $\tan 156° \ 42' \ 13'' = -0.43059$.

4. $\cot 27° \ 51' \ 17'' = 1.8923$.

5. $\cos 83° \ 17' \ 38'' = 0.11678$.

6. $\sin 87° \ 37' \ 25'' = 0.99914$.

7. $\cot 13° \ 14' \ 52'' = 4.2475$.

8. $\tan 83° \ 40' \ 30'' = 9.0218$.

Find the values of θ less than 360° in the following:

9. $\sin \theta = 0.89742$. *Ans.* 63° 49′ 12″ and 116° 10′ 48″.

10. $\cos \theta = 0.43750$. *Ans.* 64° 3′ 20″ and 295° 56′ 40″.

11. $\tan \theta = -0.92834$. *Ans.* 137° 7′ 41″ and 317° 7′ 41″.

12. $\cot \theta = 1.8923$. *Ans.* 27° 51′ 17″ and 207° 51′ 17″

13. $\cos \theta = 0.95140$. *Ans.* 17° 56′ 14″ and 342° 3′ 46″.

14. $\sin \theta = 0.13552$. *Ans.* 7° 47′ 19″ and 172° 12′ 41″.

TABLE I

FIVE-PLACE TABLE OF COMMON LOGARITHMS OF NUMBERS

From 1 to 10,000

TABLE I

FIVE-PLACE TABLE OF COMMON LOGARITHMS OF NUMBERS

From 1 to 10,000

N.	Log.	N.	Log.	N.	Log.	N.	Log.	N.	Log.
0	—	**20**	1.30 103	**40**	1.60 206	**60**	1.77 815̇	**80**	1.90 309
1	0.00 000	21	1.32 222	41	1.61 278̲	61	1.78 533	81	1.90 849
2	0.30 103	22	1.34 242	42	1.62 325̄	62	1.79 239	82	1.91 381
3	0.47 712	23	1.36 173	43	1.63 347	63	1.79 934	83	1.91 908
4	0.60 206	24	1.38 021	44	1.64 345̇	64	1.80 618	84	1.92 428
5	0.69 897	25	1.39 794	45	1.65 321	65	1.81 291	85	1.92 942
6	0.77 815	26	1.41 497	46	1.66 276	66	1.81 954	86	1.93 45̄0
7	0.84 510	27	1.43 136	47	1.67 210	67	1.82 607	87	1.93 952
8	0.90 309	28	1.44 716	48	1.68 124	68	1.83 251̲	88	1.94 448
9	0.95 424	29	1.46 240	49	1.69 020	69	1.83 885̄	89	1.94 939
10	1.00 000	**30**	1.47 712	**50**	1.69 897	**70**	1.84 510	**90**	1.95 424
11	1.04 139	31	1.49 136̲	51	1.70 757	71	1.85 126	91	1.95 904
12	1.07 918	32	1.50 515̄	52	1.71 600	72	1.85 733	92	1.96 379
13	1.11 394	33	1.51 851	53	1.72 428	73	1.86 332	93	1.96 848
14	1.14 613	34	1.53 148	54	1.73 239	74	1.86 923	94	1.97 313
15	1.17 609	35	1.54 407	55	1.74 036	75	1.87 506	95	1.97 772
16	1.20 412	36	1.55 630	56	1.74 819	76	1.88 081	96	1.98 227
17	1.23 045̄	37	1.56 820	57	1.75 587	77	1.88 649	97	1.98 677
18	1.25 527̲	38	1.57 978	58	1.76 343̲	78	1.89 209	98	1.99 123
19	1.27 875̇	39	1.59 106	59	1.77 085̇	79	1.89 763	99	1.99 564
20	1.30 103	**40**	1.60 206	**60**	1.77 815̇	**80**	1.90 309	**100**	2.00 000

N.	L. 0	1	2	3	4	5	6	7	8	9
0		00 000	30 103	47 712	60 206	69 897	77 815	84 510	90 309	95 424
1	00 000	04 139	07 918	11 394	14 613	17 609	20 412	23 045	25 527	27 875
2	30 103	32 222	34 242	36 173	38 021	39 794	41 497	43 136	44 716	46 240
3	47 712	49 136	50 515	51 851	53 148	54 407	55 630	56 820	57 978	59 106
4	60 206	61 278	62 325	63 347	64 345	65 321	66 276	67 210	68 124	69 020
5	69 897	70 757	71 600	72 428	73 239	74 036	74 819	75 587	76 343	77 085
6	77 815	78 533	79 239	79 934	80 618	81 291	81 954	82 607	83 251	83 885
7	84 510	85 126	85 733	86 332	86 923	87 506	88 081	88 649	89 209	89 763
8	90 309	90 849	91 381	91 908	92 428	92 942	93 450	93 952	94 448	94 939
9	95 424	95 904	96 379	96 848	97 313	97 772	98 227	98 677	99 123	99 564
10	00 000	00 432	00 860	01 284	01 703	02 119	02 531	02 938	03 342	03 743
11	04 139	04 532	04 922	05 308	05 690	06 070	06 446	06 819	07 188	07 555
12	07 918	08 279	08 636	08 991	09 342	09 691	10 037	10 380	10 721	11 059
13	11 394	11 727	12 057	12 385	12 710	13 033	13 354	13 672	13 988	14 301
14	14 613	14 922	15 229	15 534	15 836	16 137	16 435	16 732	17 026	17 319
15	17 609	17 898	18 184	18 469	18 752	19 033	19 312	19 590	19 866	20 140
16	20 412	20 683	20 952	21 219	21 484	21 748	22 011	22 272	22 531	22 789
17	23 045	23 300	23 553	23 805	24 055	24 304	24 551	24 797	25 042	25 285
18	25 527	25 768	26 007	26 245	26 482	26 717	26 951	27 184	27 416	27 646
19	27 875	28 103	28 330	28 556	28 780	29 003	29 226	29 447	29 667	29 885
20	30 103	30 320	30 535	30 750	30 963	31 175	31 387	31 597	31 806	32 015
21	32 222	32 428	32 634	32 838	33 041	33 244	33 445	33 646	33 846	34 044
22	34 242	34 439	34 635	34 830	35 025	35 218	35 411	35 603	35 793	35 984
23	36 173	36 361	36 549	36 736	36 922	37 107	37 291	37 475	37 658	37 840
24	38 021	38 202	38 382	38 561	38 739	38 917	39 094	39 270	39 445	39 620
25	39 794	39 967	40 140	40 312	40 483	40 654	40 824	40 993	41 162	41 330
26	41 497	41 664	41 830	41 996	42 160	42 325	42 488	42 651	42 813	42 975
27	43 136	43 297	43 457	43 616	43 775	43 933	44 091	44 248	44 404	44 560
28	44 716	44 871	45 025	45 179	45 332	45 484	45 637	45 788	45 939	46 090
29	46 240	46 389	46 538	46 687	46 835	46 982	47 129	47 276	47 422	47 567
30	47 712	47 857	48 001	48 144	48 287	48 430	48 572	48 714	48 855	48 996
31	49 136	49 276	49 415	49 554	49 693	49 831	49 969	50 106	50 243	50 379
32	50 515	50 651	50 786	50 920	51 055	51 188	51 322	51 455	51 587	51 720
33	51 851	51 983	52 114	52 244	52 375	52 504	52 634	52 763	52 892	53 020
34	53 148	53 275	53 403	53 529	53 656	53 782	53 908	54 033	54 158	54 283
35	54 407	54 531	54 654	54 777	54 900	55 023	55 145	55 267	55 388	55 509
36	55 630	55 751	55 871	55 991	56 110	56 229	56 348	56 467	56 585	56 703
37	56 820	56 937	57 054	57 171	57 287	57 403	57 519	57 634	57 749	57 864
38	57 978	58 092	58 206	58 320	58 433	58 546	58 659	58 771	58 883	58 995
39	59 106	59 218	59 329	59 439	59 550	59 660	59 770	59 879	59 988	60 097
40	60 206	60 314	60 423	60 531	60 638	60 746	60 853	60 959	61 066	61 172
41	61 278	61 384	61 490	61 595	61 700	61 805	61 909	62 014	62 118	62 221
42	62 325	62 428	62 531	62 634	62 737	62 839	62 941	63 043	63 144	63 246
43	63 347	63 448	63 548	63 649	63 749	63 849	63 949	64 048	64 147	64 246
44	64 345	64 444	64 542	64 640	64 738	64 836	64 933	65 031	65 128	65 225
45	65 321	65 418	65 514	65 610	65 706	65 801	65 896	65 992	66 087	66 181
46	66 276	66 370	66 464	66 558	66 652	66 745	66 839	66 932	67 025	67 117
47	67 210	67 302	67 394	67 486	67 578	67 669	67 761	67 852	67 943	68 034
48	68 124	68 215	68 305	68 395	68 485	68 574	68 664	68 753	68 842	68 931
49	69 020	69 108	69 197	69 285	69 373	69 461	69 548	69 636	69 723	69 810
50	69 897	69 984	70 070	70 157	70 243	70 329	70 415	70 501	70 586	70 672
N.	L. 0	1	2	3	4	5	6	7	8	9

50-100

N.	L. 0	1	2	3	4	5	6	7	8	9
50	69 897	69 984	70 070	70 157	70 243	70 329	70 415	70 501	70 586	70 672
51	70 757	70 842	70 927	71 012	71 096	71 181	71 265	71 349	71 433	71 517
52	71 600	71 684	71 767	71 850	71 933	72 016	72 099	72 181	72 263	72 346
53	72 428	72 509	72 591	72 673	72 754	72 835	72 916	72 997	73 078	73 159
54	73 239	73 320	73 400	73 480	73 560	73 640	73 719	73 799	73 878	73 957
55	74 036	74 115	74 194	74 273	74 351	74 429	74 507	74 586	74 663	74 741
56	74 819	74 896	74 974	75 051	75 128	75 205	75 282	75 358	75 435	75 511
57	75 587	75 664	75 740	75 815	75 891	75 967	76 042	76 118	76 193	76 268
58	76 343	76 418	76 492	76 567	76 641	76 716	76 790	76 864	76 938	77 012
59	77 085	77 159	77 232	77 305	77 379	77 452	77 525	77 597	77 670	77 743
60	77 815	77 887	77 960	78 032	78 104	78 176	78 247	78 319	78 390	78 462
61	78 533	78 604	78 675	78 746	78 817	78 888	78 958	79 029	79 099	79 169
62	79 239	79 309	79 379	79 449	79 518	79 588	79 657	79 727	79 796	79 865
63	79 934	80 003	80 072	80 140	80 209	80 277	80 346	80 414	80 482	80 550
64	80 618	80 686	80 754	80 821	80 889	80 956	81 023	81 090	81 158	81 224
65	81 291	81 358	81 425	81 491	81 558	81 624	81 690	81 757	81 823	81 889
66	81 954	82 020	82 086	82 151	82 217	82 282	82 347	82 413	82 478	82 543
67	82 607	82 672	82 737	82 802	82 866	82 930	82 995	83 059	83 123	83 187
68	83 251	83 315	83 378	83 442	83 506	83 569	83 632	83 696	83 759	83 822
69	83 885	83 948	84 011	84 073	84 136	84 198	84 261	84 323	84 386	84 448
70	84 510	84 572	84 634	84 696	84 757	84 819	84 880	84 942	85 003	85 065
71	85 126	85 187	85 248	85 309	85 370	85 431	85 491	85 552	85 612	85 673
72	85 733	85 794	85 854	85 914	85 974	86 034	86 094	86 153	86 213	86 273
73	86 332	86 392	86 451	86 510	86 570	86 629	86 688	86 747	86 806	86 864
74	86 923	86 982	87 040	87 099	87 157	87 216	87 274	87 332	87 390	87 448
75	87 506	87 564	87 622	87 679	87 737	87 795	87 852	87 910	87 967	88 024
76	88 081	88 138	88 195	88 252	88 309	88 366	88 423	88 480	88 536	88 593
77	88 649	88 705	88 762	88 818	88 874	88 930	88 986	89 042	89 098	89 154
78	89 209	89 265	89 321	89 376	89 432	89 487	89 542	89 597	89 653	89 708
79	89 763	89 818	89 873	89 927	89 982	90 037	90 091	90 146	90 200	90 255
80	90 309	90 363	90 417	90 472	90 526	90 580	90 634	90 687	90 741	90 795
81	90 849	90 902	90 956	91 009	91 062	91 116	91 169	91 222	91 275	91 328
82	91 381	91 434	91 487	91 540	91 593	91 645	91 698	91 751	91 803	91 855
83	91 908	91 960	92 012	92 065	92 117	92 169	92 221	92 273	92 324	92 376
84	92 428	92 480	92 531	92 583	92 634	92 686	92 737	92 788	92 840	92 891
85	92 942	92 993	93 044	93 095	93 146	93 197	93 247	93 298	93 349	93 399
86	93 450	93 500	93 551	93 601	93 651	93 702	93 752	93 802	93 852	93 902
87	93 952	94 002	94 052	94 101	94 151	94 201	94 250	94 300	94 349	94 399
88	94 448	94 498	94 547	94 596	94 645	94 694	94 743	94 792	94 841	94 890
89	94 939	94 988	95 036	95 085	95 134	95 182	95 231	95 279	95 328	95 376
90	95 424	95 472	95 521	95 569	95 617	95 665	95 713	95 761	95 809	95 856
91	95 904	95 952	95 999	96 047	96 095	96 142	96 190	96 237	96 284	96 332
92	96 379	96 426	96 473	96 520	96 567	96 614	96 661	96 708	96 755	96 802
93	96 848	96 895	96 942	96 988	97 035	97 081	97 128	97 174	97 220	97 267
94	97 313	97 359	97 405	97 451	97 497	97 543	97 589	97 635	97 681	97 727
95	97 772	97 818	97 864	97 909	97 955	98 000	98 046	98 091	98 137	98 182
96	98 227	98 272	98 318	98 363	98 408	98 453	98 498	98 543	98 588	98 632
97	98 677	98 722	98 767	98 811	98 856	98 900	98 945	98 989	99 034	99 078
98	99 123	99 167	99 211	99 255	99 300	99 344	99 388	99 432	99 476	99 520
99	99 564	99 607	99 651	99 695	99 739	99 782	99 826	99 870	99 913	99 957
100	00 000	00 043	00 087	00 130	00 173	00 217	00 260	00 303	00 346	00 389
N.	L. 0	1	2	3	4	5	6	7	8	9

TABLE I

100-150

N.	L. 0	1	2	3	4	5	6	7	8	9
100	00 000	043	087	130	173	217	260	303	346	389
101	432	475	518	561	604	647	689	732	775	817
102	860	903	945	988	*030	*072	*115	*157	*199	*242
103	01 284	326	368	410	452	494	536	578	620	662
104	703	745	787	828	870	912	953	995	*036	*078
105	02 119	160	202	243	284	325	366	407	449	490
106	531	572	612	653	694	735	776	816	857	898
107	938	979	*019	*060	*100	*141	*181	*222	*262	*302
108	03 342	383	423	463	503	543	583	623	663	703
109	743	782	822	862	902	941	981	*021	*060	*100
110	04 139	179	218	258	297	336	376	415	454	493
111	532	571	610	650	689	727	766	805	844	883
112	922	961	999	*038	*077	*115	*154	*192	*231	*269
113	05 308	346	385	423	461	500	538	576	614	652
114	690	729	767	805	843	881	918	956	994	*032
115	06 070	108	145	183	221	258	296	333	371	408
116	446	483	521	558	595	633	670	707	744	781
117	819	856	893	930	967	*004	*041	*078	*115	*151
118	07 188	225	262	298	335	372	408	445	482	518
119	555	591	628	664	700	737	773	809	846	882
120	918	954	990	*027	*063	*099	*135	*171	*207	*243
121	08 279	314	350	386	422	458	493	529	565	600
122	636	672	707	743	778	814	849	884	920	955
123	991	*026	*061	*096	*132	*167	*202	*237	*272	*307
124	09 342	377	412	447	482	517	552	587	621	656
125	691	726	760	795	830	864	899	934	968	*003
126	10 037	072	106	140	175	209	243	278	312	346
127	380	415	449	483	517	551	585	619	653	687
128	721	755	789	823	857	890	924	958	992	*025
129	11 059	093	126	160	193	227	261	294	327	361
130	394	428	461	494	528	561	594	628	661	694
131	727	760	793	826	860	893	926	959	992	*024
132	12 057	090	123	156	189	222	254	287	320	352
133	385	418	450	483	516	548	581	613	646	678
134	710	743	775	808	840	872	905	937	969	*001
135	13 033	066	098	130	162	194	226	258	290	322
136	354	386	418	450	481	513	545	577	609	640
137	672	704	735	767	799	830	862	893	925	956
138	988	*019	*051	*082	*114	*145	*176	*208	*239	*270
139	14 301	333	364	395	426	457	489	520	551	582
140	613	644	675	706	737	768	799	829	860	891
141	922	953	983	*014	*045	*076	*106	*137	*168	*198
142	15 229	259	290	320	351	381	412	442	473	503
143	534	564	594	625	655	685	715	746	776	806
144	836	866	897	927	957	987	*017	*047	*077	*107
145	16 137	167	197	227	256	286	316	346	376	406
146	435	465	495	524	554	584	613	643	673	702
147	732	761	791	820	850	879	909	938	967	997
148	17 026	056	085	114	143	173	202	231	260	289
149	319	348	377	406	435	464	493	522	551	580
150	609	638	667	696	725	754	782	811	840	869
N.	L. 0	1	2	3	4	5	6	7	8	9

Prop. Parts

	44	**43**	**42**
1	**4.4**	4.3	4.2
2	**8.8**	8.6	8.4
3	**13.2**	**12.9**	12.6
4	17.6	17.2	16.8
5	22.0	21.5	21.0
6	26.4	25.8	25.2
7	30.8	30.1	29.4
8	35.2	34.4	33.6
9	39.6	38.7	37.8
	41	**40**	**39**
1	4.1	4.0	3.9
2	8.2	8.0	7.8
3	12.3	12.0	11.7
4	16.4	16.0	15.6
5	20.5	20.0	19.5
6	24.6	24.0	23.4
7	28.7	28.0	27.3
8	32.8	32.0	31.2
9	36.9	36.0	35.1
	38	**37**	**36**
1	3.8	3.7	3.6
2	7.6	7.4	7.2
3	11.4	11.1	10.8
4	15.2	14.8	14.4
5	19.0	18.5	18.0
6	22.8	22.2	21.6
7	26.6	25.9	25.2
8	30.4	29.6	28.8
9	34.2	33.3	32.4
	35	**34**	**33**
1	3.5	3.4	3.3
2	7.0	6.8	6.6
3	10.5	10.2	9.9
4	14.0	13.6	13.2
5	17.5	17.0	16.5
6	21.0	20.4	19.8
7	24.5	23.8	23.1
8	28.0	27.2	26.4
9	31.5	30.6	29.7
	32	**31**	**30**
1	3.2	3.1	3.0
2	6.4	6.2	6.0
3	9.6	9.3	9.0
4	12.8	12.4	12.0
5	16.0	15.5	15.0
6	19.2	18.6	18.0
7	22.4	21.7	21.0
8	25.6	24.8	24.0
9	28.8	27.9	27.0

Prop. Parts

N.	L. 0	1	2	3	4	5	6	7	8	9
150	17 609	638	667	696	725	754	782	811	840	869
151	898	926	955	984	*013	*041	*070	*099	*127	*156
152	18 184	213	241	270	298	327	355	384	412	441
153	469	498	526	554	583	611	639	667	696	724
154	752	780	808	837	865	893	921	949	977	*005
155	19 033	061	089	117	145	173	201	229	257	285
156	312	340	368	396	424	451	479	507	535	562
157	590	618	645	673	700	728	756	783	811	838
158	866	893	921	948	976	*003	*030	*058	*085	*112
159	20 140	167	194	222	249	276	303	330	358	385
160	412	439	466	493	520	548	575	602	629	656
161	683	710	737	763	790	817	844	871	898	925
162	952	978	*005	*032	*059	*085	*112	*139	*165	*192
163	21 219	245	272	299	325	352	378	405	431	458
164	484	511	537	564	590	617	643	669	696	722
165	748	775	801	827	854	880	906	932	958	985
166	22 011	037	063	089	115	141	167	194	220	246
167	272	298	324	350	376	401	427	453	479	505
168	531	557	583	608	634	660	686	712	737	763
169	789	814	840	866	891	917	943	968	994	*019
170	23 045	070	096	121	147	172	198	223	249	274
171	300	325	350	376	401	426	452	477	502	528
172	553	578	603	629	654	679	704	729	754	779
173	805	830	855	880	905	930	955	980	*005	*030
174	24 055	080	105	130	155	180	204	229	254	279
175	304	329	353	378	403	428	452	477	502	527
176	551	576	601	625	650	674	699	724	748	773
177	797	822	846	871	895	920	944	969	993	*018
178	25 042	066	091	115	139	164	188	212	237	261
179	285	310	334	358	382	406	431	455	479	503
180	527	551	575	600	624	648	672	696	720	744
181	768	792	816	840	864	888	912	935	959	983
182	26 007	031	055	079	102	126	150	174	198	221
183	245	269	293	316	340	364	387	411	435	458
184	482	505	529	553	576	600	623	647	670	694
185	717	741	764	788	811	834	858	881	905	928
186	951	975	998	*021	*045	*068	*091	*114	*138	*161
187	27 184	207	231	254	277	300	323	346	370	393
188	416	439	462	485	508	531	554	577	600	623
189	646	669	692	715	738	761	784	807	830	852
190	875	898	921	944	967	989	*012	*035	*058	*081
191	28 103	126	149	171	194	217	240	262	285	307
192	330	353	375	398	421	443	466	488	511	533
193	556	578	601	623	646	668	691	713	735	758
194	780	803	825	847	870	892	914	937	959	981
195	29 003	026	048	070	092	115	137	159	181	203
196	226	248	270	292	314	336	358	380	403	425
197	447	469	491	513	535	557	579	601	623	645
198	667	688	710	732	754	776	798	820	842	863
199	885	907	929	951	973	994	*016	*038	*060	*081
200	30 103	125	146	168	190	211	233	255	276	298
N.	L. 0	1	2	3	4	5	6	7	8	9

Prop. Parts

	29	28	27	26	25	24	23	22	21
1	2.9	2.8	2.7	2.6	2.5	2.4	2.3	2.2	2.1
2	5.8	5.6	5.4	5.2	5.0	4.8	4.6	4.4	4.2
3	8.7	8.4	8.1	7.8	7.5	7.2	6.9	6.6	6.3
4	11.6	11.2	10.8	10.4	10.0	9.6	9.2	8.8	8.4
5	14.5	14.0	13.5	13.0	12.5	12.0	11.5	11.0	10.5
6	17.4	16.8	16.2	15.6	15.0	14.4	13.8	13.2	12.6
7	20.3	19.6	18.9	18.2	17.5	16.8	16.1	15.4	14.7
8	23.2	22.4	21.6	20.8	20.0	19.2	18.4	17.6	16.8
9	26.1	25.2	24.3	23.4	22.5	21.6	20.7	19.8	18.9

Prop. Parts

N.	L. 0	1	2	3	4	5	6	7	8	9
200	30 103	125̄	146	168	190	211	233	255̄	276	298
201	320	341	363	384	406	428	449	471	492	514
202	535	557	578	600	621	643	664	685	707	728
203	750̄	771	792	814	835	856	878	899	920	942
204	963	984	*006	*027	*048	*069	*091	*112	*133	*154
205	31 175	197	218	239	260	281	302	323	345̄	366
206	387	408	429	450̄	471	492	513	534	555	576
207	597	618	639	660	681	702	723	744	765̄	785
208	806	827	848	869	890	911	931	952	973	994
209	32 015̄	035	056	077	098	118	139	160	181	201
210	222	243	263	284	305̄	325	346	366	387	408
211	428	449	469	490	510	531	552	572	593	613
212	634	654	675̄	695̄	715	736	756	777	797	818
213	838	858	879	899	919	940	960	980	*001	*021
214	33 041	062	082	102	122	143	163	183	203	224
215	244	264	284	304	325̄	345̄	365̄	385	405	425
216	445	465	486	506	526	546	566	586	606	626
217	646	666	686	706	726	746	766	786	806	826
218	846	866	885	905	925	945	965	985̄	*005̄	*025̄
219	34 044	064	084	104	124	143	163	183	203	223
220	242	262	282	301	321	341	361	380	400	420
221	439	459	479	498	518	537	557	577	596	616
222	635	655̄	674	694	713	733	753	772	792	811
223	830	850̄	869	889	908	928	947	967	986	*005
224	35 025̄	044	064	083	102	122	141	160	180	199
225	218	238	257	276	295	315̄	334	353	372	392
226	411	430	449	468	488	507	526	545	564	583
227	603	622	641	660	679	698	717	736	755	774
228	793	813	832	851	870	889	908	927	946	965̄
229	984	*003	*021	*040	*059	*078	*097	*116	*135	*154
230	36 173	192	211	229	248	267	286	305̄	324	342
231	361	380	399	418	436	455	474	493	511	530
232	549	568	586	605̄	624	642	661	680	698	717
233	736	754	773	791	810	829	847	866	884	903
234	922	940	959	977	996	*014	*033	*051	*070	*088
235	37 107	125	144	162	181	199	218	236	254	273
236	291	310	328	346	365̄	383	401	420	438	457
237	475̄	493	511	530	548	566	585̄	603	621	639
238	658	676	694	712	731	749	767	785	803	822
239	840	858	876	894	912	931	949	967	985̄	*003
240	38 021	039	057	075	093	112	130	148	166	184
241	202	220	238	256	274	292	310	328	346	364
242	382	399	417	435	453	471	489	507	525̄	543
243	561	578	596	614	632	650̄	668	686	703	721
244	739	757	775̄	792	810	828	846	863	881	899
245	917	934	952	970	987	*005	*023	*041	*058	*076
246	39 094	111	129	146	164	182	199	217	235̄	252
247	270	287	305̄	322	340	358	375	393	410	428
248	445	463	480	498	515	533	550	568	585	602
249	620	637	655̄	672	690	707	724	742	759	777
250	794	811	829	846	863	881	898	915	933	950
N.	L. 0	1	2	3	4	5	6	7	8	9

Prop. Parts

	22	21
1	2.2	2.1
2	4.4	4.2
3	6.6	6.3
4	8.8	8.4
5	11.0	10.5
6	13.2	12.6
7	15.4	14.7
8	17.6	16.8
9	19.8	18.9

	20
1	2.0
2	4.0
3	6.0
4	8.0
5	10.0
6	12.0
7	14.0
8	16.0
9	18.0

	19
1	1.9
2	3.8
3	5.7
4	7.6
5	9.5
6	11.4
7	13.3
8	15.2
9	17.1

	18
1	1.8
2	3.6
3	5.4
4	7.2
5	9.0
6	10.8
7	12.6
8	14.4
9	16.2

	17
1	1.7
2	3.4
3	5.1
4	6.8
5	8.5
6	10.2
7	11.9
8	13.6
9	15.3

Prop. Parts

250-300

N.	L. 0	1	2	3	4	5	6	7	8	9
250	39 794	811	829	846	863	881	898	915	933	950
251	967	985̄	*002	*019	*037	*054	*071	*088	*106	*123
252	40 140	157	175̄	192	209	226	243	261	278	295̄
253	312	329	346	364	381	398	415̄	432	449	466
254	483	500	518	535̄	552	569	586	603	620	637
255	654	671	688	705	722	739	756	773	790	807
256	824	841	858	875̄	892	909	926	943	960	976
257	993	*010	*027	*044	*061	*078	*095̄	*111	*128	*145
258	41 162	179	196	212	229	246	263	280	296	313
259	330	347	363	380	397	414	430	447	464	481
260	497	514	531	547	564	581	597	614	631	647
261	664	681	697	714	731	747	764	780	797	814
262	830	847	863	880	896	913	929	946	963	979
263	996	*012	*029	*045	*062	*078	*095̄	*111	*127	*144
264	42 160	177	193	210	226	243	259	275	292	308
265	325̄	341	357	374	390	406	423	439	455	472
266	488	504	521	537	553	570	586	602	619	635̄
267	651	667	684	700	716	732	749	765̄	781	797
268	813	830	846	862	878	894	911	927	943	959
269	975	991	*008	*024	*040	*056	*072	*088	*104	*120
270	43 136	152	169	185̄	201	217	233	249	265̄	281
271	297	313	329	345̄	361	377	393	409	425̄	441
272	457	473	489	505̄	521	537	553	569	584	600
273	616	632	648	664	680	696	712	727	743	759
274	775	791	807	823	838	854	870	886	902	917
275	933	949	965̄	981	996	*012	*028	*044	*059	*075
276	44 091	107	122	138	154	170	185	201	217	232
277	248	264	279	295̄	311	326	342	358	373	389
278	404	420	436	451	467	483	498	514	529	545̄
279	560	576	592	607	623	638	654	669	685̄	700
280	716	731	747	762	778	793	809	824	840	855
281	871	886	902	917	932	948	963	979	994	*010
282	45 025̄	040	056	071	086	102	117	133	148	163
283	179	194	209	225̄	240	255	271	286	301	317
284	332	347	362	378	393	408	423	439	454	469
285	484	5̄00	515̄	530	545	561	576	591	606	621
286	637	652	667	682	697	712	728	743	758	773
287	788	803	818	834	849	864	879	894	909	924
288	939	954	969	984	*000	*015̄	*030	*045̄	*060	*075̄
289	46 090	105̄	120	135̄	15̄0	165̄	180	195̄	210	225̄
290	240	255̄	270	285̄	300	315̄	330	345̄	359	374
291	389	404	419	434	449	464	479	494	509	523
292	538	553	568	583	598	613	627	642	657	672
293	687	702	716	731	746	761	776	790	805	820
294	835̄	85̄0	864	879	894	909	923	938	953	967
295	982	997	*012	*026	*041	*056	*070	*085	*100	*114
296	47 129	144	159	173	188	202	217	232	246	261
297	276	290	305̄	319	334	349	363	378	392	407
298	422	436	451	465	480	494	509	524	538	553
299	567	582	596	611	625	640	654	669	683	698
300	712	727	741	756	770	784	799	813	828	842
N.	L. 0	1	2	3	4	5	6	7	8	9

Prop. Parts

	18	17	16	15	14
1	1.8	1.7	1.6	1.5	1.4
2	3.6	3.4	3.2	3.0	2.8
3	5.4	5.1	4.8	4.5	4.2
4	7.2	6.8	6.4	6.0	5.6
5	9.0	8.5	8.0	7.5	7.0
6	10.8	10.2	9.6	9.0	8.4
7	12.6	11.9	11.2	10.5	9.8
8	14.4	13.6	12.8	12.0	11.2
9	16.2	15.3	14.4	13.5	12.6

$\log e = 0.43429$

N.	L. 0	1	2	3	4	5	6	7	8	9
300	47 712	727	741	756	770	784	799	813	828	842
301	857	871	885	900	914	929	943	958	972	986
302	48 001	015	029	044	058	073	087	101	116	130
303	144	159	173	187	202	216	230	244	259	273
304	287	302	316	330	344	359	373	387	401	416
305	430	444	458	473	487	501	515	530	544	558
306	572	586	601	615̄	629	643	657	671	686	700
307	714	728	742	756	770	785̄	799	813	827	841
308	855	869	883	897	911	926	940	954	968	982
309	996	*010	*024	*038	*052	*066	*080	*094	*108	*122
310	49 136	150	164	178	192	206	220	234	248	262
311	276	290	304	318	332	346	360	374	388	402
312	415	429	443	457	471	485	499	513	527	541
313	554	568	582	596	610	624	638	651	665	679
314	693	707	721	734	748	762	776	790	803	817
315	831	845̄	859	872	886	900	914	927	941	955̄
316	969	982	996	*010	*024	*037	*051	*065̄	*079	*092
317	50 106	120	133	147	161	174	188	202	215	229
318	243	256	270	284	297	311	325̄	338	352	365
319	379	393	406	420	433	447	461	474	488	501
320	515̄	529	542	556	569	583	596	610	623	637
321	651	664	678	691	705̄	718	732	745	759	772
322	786	799	813	826	840	853	866	880	893	907
323	920	934	947	961	974	987	*001	*014	*028	*041
324	51 055̄	068	081	095̄	108	121	135̄	148	162	175̄
325	188	202	215	228	242	255	268	282	295	308
326	322	335	348	362	375	388	402	415̄	428	441
327	455̄	468	481	495̄	508	521	534	548	561	574
328	587	601	614	627	640	654	667	680	693	706
329	720	733	746	759	772	786	799	812	825	838
330	851	865̄	878	891	904	917	930	943	957	970
331	983	996	*009	*022	*035	*048	*061	*075̄	*088	*101
332	52 114	127	140	153	166	179	192	205	218	231
333	244	257	270	284	297	310	323	336	349	362
334	375̄	388	401	414	427	440	453	466	479	492
335	504	517	530	543	556	569	582	595	608	621
336	634	647	660	673	686	699	711	724	737	750
337	763	776	789	802	815̄	827	840	853	866	879
338	892	905̄	917	930	943	956	969	982	994	*007
339	53 020	033	046	058	071	084	097	110	122	135
340	148	161	173	186	199	212	224	237	25̄0	263
341	275	288	301	314	326	339	352	364	377	390
342	403	415	428	441	453	466	479	491	504	517
343	529	542	555̄	567	580	593	605	618	631	643
344	656	668	681	694	706	719	732	744	757	769
345	782	794	807	820	832	845̄	857	870	882	895
346	908	920	933	945	958	970	983	995	*008	*020
347	54 033	045	058	070	083	095	108	120	133	145
348	158	170	183	195	208	220	233	245	258	270
349	283	295̄	307	320	332	345̄	357	370	382	394
350	407	419	432	444	456	469	481	494	506	518
N.	L. 0	1	2	3	4	5	6	7	8	9

Prop. Parts

	15	14	13	12
1	1.5	1.4	1.3	1.2
2	3.0	2.8	2.6	2.4
3	4.5	4.2	3.9	3.6
4	6.0	5.6	5.2	4.8
5	7.5	7.0	6.5	6.0
6	9.0	8.4	7.8	7.2
7	10.5	9.8	9.1	8.4
8	12.0	11.2	10.4	9.6
9	13.5	12.6	11.7	10.8

$\log \pi = 0.49715$

350-400

N.	L. 0	1	2	3	4	5	6	7	8	9
350	54 407	419	432	444	456	469	481	494	506	518
351	531	543	555	568	580	593	60$\bar{5}$	617	630	642
352	654	667	679	691	704	716	728	741	753	765
353	777	790	802	814	827	839	851	864	876	888
354	900	913	92$\bar{5}$	937	949	962	974	986	998	*011
355	55 023	035	047	060	072	084	096	108	121	133
356	14$\bar{5}$	157	169	182	194	206	218	230	242	25$\bar{5}$
357	267	279	291	303	315	328	340	352	364	376
358	388	400	413	42$\bar{5}$	437	449	461	473	485	497
359	509	522	534	546	558	570	582	594	606	618
360	630	642	654	666	678	691	703	71$\bar{5}$	727	739
361	751	763	77$\bar{5}$	787	799	811	823	83$\bar{5}$	847	859
362	871	883	89$\bar{5}$	907	919	931	943	95$\bar{5}$	967	979
363	991	*003	*01$\bar{5}$	*027	*038	*050	*062	*074	*086	*098
364	56 110	122	134	146	158	170	182	194	205	217
365	229	241	253	26$\bar{5}$	277	289	301	312	324	336
366	348	360	372	384	396	407	419	431	443	45$\bar{5}$
367	467	478	490	502	514	526	538	549	561	573
368	58$\bar{5}$	597	608	620	632	644	656	667	679	691
369	703	714	726	738	75$\bar{0}$	761	773	78$\bar{5}$	797	808
370	820	832	844	855	867	879	891	902	914	926
371	937	949	961	972	984	996	*008	*019	*031	*043
372	57 054	066	078	089	101	113	124	136	148	159
373	171	183	194	206	217	229	241	252	264	276
374	287	299	310	322	334	345	357	368	380	392
375	403	41$\bar{5}$	426	438	449	461	473	484	496	507
376	519	530	542	553	56$\bar{5}$	576	588	600	611	623
377	634	646	657	669	680	692	703	71$\bar{5}$	726	738
378	749	761	772	784	795	807	818	830	841	852
379	864	875	887	898	910	921	933	944	955	967
380	978	990	*001	*013	*024	*035	*047	*058	*070	*081
381	58 092	104	115	127	138	149	161	172	184	19$\bar{5}$
382	206	218	229	240	252	263	274	286	297	309
383	320	331	343	354	365	377	388	399	410	422
384	433	444	456	467	478	490	501	512	524	53$\bar{5}$
385	546	557	569	580	591	602	614	62$\bar{5}$	636	647
386	659	670	681	692	704	71$\bar{5}$	726	737	749	760
387	771	782	794	80$\bar{5}$	816	827	838	85$\bar{0}$	861	872
388	883	894	906	917	928	939	950	961	973	984
389	99$\bar{5}$	*006	*017	*028	*040	*051	*062	*073	*084	*095
390	59 106	118	129	140	151	162	173	184	195	207
391	218	229	240	251	262	273	284	295	306	318
392	329	340	351	362	373	384	395	406	417	428
393	439	450	461	472	483	494	506	517	528	539
394	55$\bar{0}$	561	572	583	594	60$\bar{5}$	616	627	638	649
395	660	671	682	693	704	71$\bar{5}$	726	737	748	759
396	770	780	791	802	813	824	835	846	857	868
397	879	890	901	912	923	934	94$\bar{5}$	956	966	977
398	988	999	*010	*021	*032	*043	*054	*06$\bar{5}$	*076	*086
399	60 097	108	119	130	141	152	163	173	184	195
400	206	217	228	239	249	260	271	282	293	304
N.	L. 0	1	2	3	4	5	6	7	8	9

Prop. Parts

	13
1	1.3
2	2.6
3	3.9
4	5.2
5	6.5
6	7.8
7	9.1
8	10.4
9	11.7

	12
1	1.2
2	2.4
3	3.6
4	4.8
5	6.0
6	7.2
7	8.4
8	9.6
9	10.8

	11
1	1.1
2	2.2
3	3.3
4	4.4
5	5.5
6	6.6
7	7.7
8	8.8
9	9.9

	10
1	1.0
2	2.0
3	3.0
4	4.0
5	5.0
6	6.0
7	7.0
8	8.0
9	9.0

400-450

N.	L. 0	1	2	3	4	5	6	7	8	9
400	60 206	217	228	239	249	260	271	282	293	304
401	314	325	336	347	358	369	379	390	401	412
402	423	433	444	455	466	477	487	498	509	520
403	531	541	552	563	574	584	595	606	617	627
404	638	649	660	670	681	692	703	713	724	735̄
405	746	756	767	778	788	799	810	821	831	842
406	853	863	874	885̄	895	906	917	927	938	949
407	959	970	981	991	*002	*013	*023	*034	*045̄	*055
408	61 066	077	087	098	109	119	130	140	151	162
409	172	183	194	204	215̄	225	236	247	257	268
410	278	289	300	310	321	331	342	352	363	374
411	384	395̄	405	416	426	437	448	458	469	479
412	490	500	511	521	532	542	553	563	574	584
413	595	606	616	627	637	648	658	669	679	690
414	700	711	721	731	742	752	763	773	784	794
415	805̄	815	826	836	847	857	868	878	888	899
416	909	920	930	941	951	962	972	982	993	*003
417	62 014	024	034	045̄	055	066	076	086	097	107
418	118	128	138	149	159	170	180	190	201	211
419	221	232	242	252	263	273	284	294	304	315̄
420	325̄	335	346	356	366	377	387	397	408	418
421	428	439	449	459	469	480	490	500	511	521
422	531	542	552	562	572	583	593	603	613	624
423	634	644	655̄	665̄	675	685	696	706	716	726
424	737	747	757	767	778	788	798	808	818	829
425	839	849	859	870	880	890	900	910	921	931
426	941	951	961	972	982	992	*002	*012	*022	*033
427	63 043	053	063	073	083	094	104	114	124	134
428	144	155̄	165̄	175̄	185̄	195	205	215	225	236
429	246	256	266	276	286	296	306	317	327	337
430	347	357	367	377	387	397	407	417	428	438
431	448	458	468	478	488	498	508	518	528	538
432	548	558	568	579	589	599	609	619	629	639
433	649	659	669	679	689	699	709	719	729	739
434	749	759	769	779	789	799	809	819	829	839
435	849	859	869	879	889	899	909	919	929	939
436	949	959	969	979	988	998	*008	*018	*028	*038
437	64 048	058	068	078	088	098	108	118	128	137
438	147	157	167	177	187	197	207	217	227	237
439	246	256	266	276	286	296	306	316	326	335
440	345	355	365	375̄	385̄	395̄	404	414	424	434
441	444	454	464	473	483	493	503	513	523	532
442	542	552	562	572	582	591	601	611	621	631
443	640	650	660	670	680	689	699	709	719	729
444	738	748	758	768	777	787	797	807	816	826
445	836	846	856	865	875	885̄	895̄	904	914	924
446	933	943	953	963	972	982	992	*002	*011	*021
447	65 031	040	050	060	070	079	089	099	108	118
448	128	137	147	157	167	176	186	196	205	215̄
449	225̄	234	244	254	263	273	283	292	302	312
450	321	331	341	350	360	369	379	389	398	408
N.	L. 0	1	2	3	4	5	6	7	8	9

Prop. Parts

	11	10	9
1	1.1	1.0	0.9
2	2.2	2.0	1.8
3	3.3	3.0	2.7
4	4.4	4.0	3.6
5	5.5	5.0	4.5
6	6.6	6.0	5.4
7	7.7	7.0	6.3
8	8.8	8.0	7.2
9	9.9	9.0	8.1

450-500

N.	L. 0	1	2	3	4	5	6	7	8	9	Prop. Parts
450	65 321	331	341	350	360	369	379	389	398	408	
451	418	427	437	447	456	466	475	485	49$\bar{5}$	504	
452	514	523	533	543	552	562	571	581	591	600	
453	610	619	629	639	648	658	667	677	686	696	
454	706	715	72$\bar{5}$	734	744	753	763	772	782	792	
455	801	811	820	830	839	849	858	868	877	887	
456	896	906	916	925	93$\bar{5}$	944	954	963	973	982	
457	992	*001	*011	*020	*030	*039	*049	*058	*068	*077	
458	66 087	096	106	11$\bar{5}$	124	134	143	153	162	172	
459	181	191	200	210	219	229	238	247	257	266	
460	276	285	29$\bar{5}$	304	314	323	332	342	351	361	
461	370	380	389	398	408	417	427	436	445	45$\bar{5}$	
462	464	474	483	492	502	511	521	530	539	549	
463	558	567	577	586	596	60$\bar{5}$	614	624	633	642	
464	652	661	671	680	689	699	708	717	727	736	
465	745	75$\bar{5}$	764	773	783	792	801	811	820	829	
466	839	848	857	867	876	885	894	904	913	922	
467	932	941	950	960	969	978	987	997	*006	*015	
468	67 02$\bar{5}$	034	043	052	062	071	080	089	099	108	
469	117	127	136	145	154	164	173	182	191	201	
470	210	219	228	237	247	256	265	274	284	293	
471	302	311	321	330	339	348	357	367	376	38$\bar{5}$	
472	394	403	413	422	431	440	449	459	468	477	
473	486	495	504	514	523	532	541	550	560	569	
474	578	587	596	605	614	624	633	642	651	660	
475	669	679	688	697	706	715	724	733	742	752	
476	761	770	779	788	797	806	815	82$\bar{5}$	834	843	
477	852	861	870	879	888	897	906	916	92$\bar{5}$	934	
478	943	952	961	970	979	988	997	*006	*015	*024	
479	68 034	043	052	061	070	079	088	097	106	115	
480	124	133	142	151	160	169	178	187	196	205	
481	21$\bar{5}$	224	233	242	251	260	269	278	287	296	
482	30$\bar{5}$	314	323	332	341	3$\bar{5}$0	359	368	377	386	
483	39$\bar{5}$	404	413	422	431	440	449	458	467	476	
484	48$\bar{5}$	494	502	511	520	529	538	547	556	565	
485	574	583	592	601	610	619	628	637	646	65$\bar{5}$	
486	664	673	681	690	699	708	717	726	735	744	
487	753	762	771	780	789	797	806	815	824	833	
488	842	851	860	869	878	886	895	904	913	922	
489	931	940	949	958	966	975	984	993	*002	*011	
490	69 020	028	037	046	055	064	073	082	090	099	
491	108	117	126	13$\bar{5}$	144	152	161	170	179	188	
492	197	205	214	223	232	241	249	258	267	276	
493	28$\bar{5}$	294	302	311	320	329	338	346	355	364	
494	373	381	390	399	408	417	425	434	443	452	
495	461	469	478	487	496	504	513	522	531	539	
496	548	557	566	574	583	592	601	609	618	627	
497	636	644	653	662	671	679	688	697	705	714	
498	723	732	740	749	758	767	775	784	793	801	
499	810	819	827	836	84$\bar{5}$	854	862	871	880	888	
500	897	906	914	923	932	940	949	958	966	975	
N.	L. 0	1	2	3	4	5	6	7	8	9	Prop. Parts

Prop. Parts

	10
1	1.0
2	2.0
3	3.0
4	4.0
5	5.0
6	6.0
7	7.0
8	8.0
9	9.0

	9
1	0.9
2	1.8
3	2.7
4	3.6
5	4.5
6	5.4
7	6.3
8	7.2
9	8.1

	8
1	0.8
2	1.6
3	2.4
4	3.2
5	4.0
6	4.8
7	5.6
8	6.4
9	7.2

500-550

N.	L. 0	1	2	3	4	5	6	7	8	9
500	69 897	906	914	923	932	940	949	958	966	975
501	984	992	*001	*010	*018	*027	*036	*044	*053	*062
502	70 070	079	088	096	105	114	122	131	140	148
503	157	165	174	183	191	200	209	217	226	234
504	243	252	260	269	278	286	295	303	312	321
505	329	338	346	355	364	372	381	389	398	406
506	415	424	432	441	449	458	467	475	484	492
507	501	509	518	526	535	544	552	561	569	578
508	586	595	603	612	621	629	638	646	655	663
509	672	680	689	697	706	714	723	731	740	749
510	757	766	774	783	791	800	808	817	825	834
511	842	851	859	868	876	885	893	902	910	919
512	927	935	944	952	961	969	978	986	995	*003
513	71 012	020	029	037	046	054	063	071	079	088
514	096	105	113	122	130	139	147	155	164	172
515	181	189	198	206	214	223	231	240	248	257
516	265	273	282	290	299	307	315	324	332	341
517	349	357	366	374	383	391	399	408	416	425
518	433	441	450	458	466	475	483	492	500	508
519	517	525	533	542	550	559	567	575	584	592
520	600	609	617	625	634	642	650	659	667	675
521	684	692	700	709	717	725	734	742	750	759
522	767	775	784	792	800	809	817	825	834	842
523	850	858	867	875	883	892	900	908	917	925
524	933	941	950	958	966	975	983	991	999	*008
525	72 016	024	032	041	049	057	066	074	082	090
526	099	107	115	123	132	140	148	156	165	173
527	181	189	198	206	214	222	230	239	247	255
528	263	272	280	288	296	304	313	321	329	337
529	346	354	362	370	378	387	395	403	411	419
530	428	436	444	452	460	469	477	485	493	501
531	509	518	526	534	542	550	558	567	575	583
532	591	599	607	616	624	632	640	648	656	665
533	673	681	689	697	705	713	722	730	738	746
534	754	762	770	779	787	795	803	811	819	827
535	835	843	852	860	868	876	884	892	900	908
536	916	925	933	941	949	957	965	973	981	989
537	997	*006	*014	*022	*030	*038	*046	*054	*062	*070
538	73 078	086	094	102	111	119	127	135	143	151
539	159	167	175	183	191	199	207	215	223	231
540	239	247	255	263	272	280	288	296	304	312
541	320	328	336	344	352	360	368	376	384	392
542	400	408	416	424	432	440	448	456	464	472
543	480	488	496	504	512	520	528	536	544	552
544	560	568	576	584	592	600	608	616	624	632
545	640	648	656	664	672	679	687	695	703	711
546	719	727	735	743	751	759	767	775	783	791
547	799	807	815	823	830	838	846	854	862	870
548	878	886	894	902	910	918	926	933	941	949
549	957	965	973	981	989	997	*005	*013	*020	*028
550	74 036	044	052	060	068	076	084	092	099	107
N.	L. 0	1	2	3	4	5	6	7	8	9

Prop. Parts

	9	8	7
1	0.9	0.8	0.7
2	1.8	1.6	1.4
3	2.7	2.4	2.1
4	3.6	3.2	2.8
5	4.5	4.0	3.5
6	5.4	4.8	4.2
7	6.3	5.6	4.9
8	7.2	6.4	5.6
9	8.1	7.2	6.3

550-600

N.	L. 0	1	2	3	4	5	6	7	8	9
550	74 036	044	052	060	068	076	084	092	099	107
551	115	123	131	139	147	$15\bar{5}$	162	170	178	186
552	194	202	210	218	225	233	241	249	257	$26\bar{5}$
553	273	280	288	296	304	312	320	327	335	343
554	351	359	367	374	382	390	398	406	414	421
555	429	437	445	453	461	468	476	484	492	$\bar{5}00$
556	507	515	523	531	539	547	554	562	570	578
557	586	593	601	609	617	624	632	640	648	656
558	663	671	679	687	$69\bar{5}$	702	710	718	726	733
559	741	749	757	764	772	780	788	796	803	811
560	819	827	834	842	$8\bar{5}0$	858	865	873	881	889
561	896	904	912	920	927	$93\bar{5}$	943	950	958	966
562	974	981	989	997	*$00\bar{5}$	*012	*020	*028	*035	*043
563	75 051	059	066	074	082	089	097	$10\bar{5}$	113	120
564	128	136	143	151	159	166	174	182	189	197
565	$20\bar{5}$	213	220	228	236	243	251	259	266	274
566	282	289	297	$30\bar{5}$	312	320	328	335	343	351
567	358	366	374	381	389	397	404	412	420	427
568	$43\bar{5}$	442	450	458	465	473	481	488	496	504
569	511	519	526	534	542	549	557	$56\bar{5}$	572	580
570	587	595	603	610	618	626	633	641	648	656
571	664	671	679	686	694	702	709	717	724	732
572	740	747	$75\bar{5}$	762	770	778	785	793	800	808
573	815	823	831	838	846	853	861	868	876	884
574	891	899	906	914	921	929	937	944	952	959
575	967	974	982	989	997	*$00\bar{5}$	*012	*020	*027	*$03\bar{5}$
576	76 042	$0\bar{5}0$	057	$06\bar{5}$	072	080	087	$09\bar{5}$	103	110
577	118	125	133	140	148	155	163	170	178	185
578	193	200	208	215	223	230	238	245	253	260
579	268	275	283	290	298	305	313	320	328	335
580	343	350	358	365	373	380	388	395	403	410
581	418	425	433	440	448	$45\bar{5}$	462	470	477	$48\bar{5}$
582	492	$\bar{5}00$	507	$51\bar{5}$	522	530	537	$54\bar{5}$	552	559
583	567	574	582	589	597	604	612	619	626	634
584	641	649	656	664	671	678	686	693	701	708
585	716	723	730	738	745	753	760	768	$77\bar{5}$	782
586	790	797	$80\bar{5}$	812	819	827	834	842	849	856
587	864	871	879	886	893	901	908	916	923	930
588	938	945	953	960	967	$97\bar{5}$	982	989	997	*004
589	77 012	019	026	034	041	048	056	063	070	078
590	085	093	100	107	$11\bar{5}$	122	129	137	144	151
591	159	166	173	181	188	195	203	210	217	$22\bar{5}$
592	232	240	247	254	262	269	276	283	291	298
593	305	313	320	327	$33\bar{5}$	342	349	357	364	371
594	379	386	393	401	408	415	422	430	437	444
595	452	459	466	474	481	488	495	503	510	517
596	525	532	539	546	554	561	568	576	583	590
597	597	$60\bar{5}$	612	619	627	634	641	648	656	663
598	670	677	$68\bar{5}$	692	699	706	714	721	728	735
599	743	$7\bar{5}0$	757	764	772	779	786	793	801	808
600	815	822	830	837	844	851	859	866	873	880
N.	L. 0	1	2	3	4	5	6	7	8	9

Prop. Parts

	8
1	0.8
2	1.6
3	2.4
4	3.2
5	4.0
6	4.8
7	5.6
8	6.4
9	7.2

	7
1	0.7
2	1.4
3	2.1
4	2.8
5	3.5
6	4.2
7	4.9
8	5.6
9	6.3

600-650

N.	L. 0	1	2	3	4	5	6	7	8	9
600	77 815	822	830	837	844	851	859	866	873	880
601	887	895̄	902	909	916	924	931	938	945	952
602	960	967	974	981	988	996	*003	*010	*017	*025̄
603	78 032	039	046	053	061	068	075̄	082	089	097
604	104	111	118	125	132	140	147	154	161	168
605	176	183	190	197	204	211	219	226	233	240
606	247	254	262	269	276	283	290	297	305̄	312
607	319	326	333	340	347	355̄	362	369	376	383
608	390	398	405̄	412	419	426	433	440	447	455̄
609	462	469	476	483	490	497	504	512	519	526
610	533	540	547	554	561	569	576	583	590	597
611	604	611	618	625	633	640	647	654	661	668
612	675	682	689	696	704	711	718	725̄	732	739
613	746	753	760	767	774	781	789	796	803	810
614	817	824	831	838	845	852	859	866	873	880
615	888	895̄	902	909	916	923	930	937	944	951
616	958	965	972	979	986	993	*000	*007	*014	*021
617	79 029	036	043	050̄	057	064	071	078	085̄	092
618	099	106	113	120	127	134	141	148	155	162
619	169	176	183	190	197	204	211	218	225	232
620	239	246	253	260	267	274	281	288	295	302
621	309	316	323	330	337	344	351	358	365	372
622	379	386	393	400	407	414	421	428	435̄	442
623	449	456	463	470	477	484	491	498	505̄	511
624	518	525	532	539	546	553	560	567	574	581
625	588	595̄	602	609	616	623	630	637	644	650
626	657	664	671	678	685	692	699	706	713	720
627	727	734	741	748	754	761	768	775	782	789
628	796	803	810	817	824	831	837	844	851	858
629	865	872	879	886	893	900	906	913	920	927
630	934	941	948	955̄	962	969	975	982	989	996
631	80 003	010	017	024	030	037	044	051	058	065̄
632	072	079	085	092	099	106	113	120	127	134
633	140	147	154	161	168	175̄	182	188	195	202
634	209	216	223	229	236	243	250	257	264	271
635	277	284	291	298	305̄	312	318	325	332	339
636	346	353	359	366	373	380	387	393	400	407
637	414	421	428	434	441	448	455̄	462	468	475
638	482	489	496	502	509	516	523	530	536	543
639	550	557	564	570	577	584	591	598	604	611
640	618	625̄	632	638	645	652	659	665	672	679
641	686	693	699	706	713	720	726	733	740	747
642	754	760	767	774	781	787	794	801	808	814
643	821	828	835̄	841	848	855̄	862	868	875	882
644	889	895	902	909	916	922	929	936	943	949
645	956	963	969	976	983	990	996	*003	*010	*017
646	81 023	030	037	043	050	057	064	070	077	084
647	090	097	104	111	117	124	131	137	144	151
648	158	164	171	178	184	191	198	204	211	218
649	224	231	238	245̄	251	258	265̄	271	278	285̄
650	291	298	305̄	311	318	325̄	331	338	345̄	351

| N. | L. 0 | 1 | 2 | 3 | 4 | 5 | 6 | 7 | 8 | 9 | Prop. Parts |

Prop. Parts

	8
1	0.8
2	1.6
3	2.4
4	3.2
5	4.0
6	4.8
7	5.6
8	6.4
9	7.2

	7
1	0.7
2	1.4
3	2.1
4	2.8
5	3.5
6	4.2
7	4.9
8	5.6
9	6.3

	6
1	0.6
2	1.2
3	1.8
4	2.4
5	3.0
6	3.6
7	4.2
8	4.8
9	5.4

650-700

N.	L. 0	1	2	3	4	5	6	7	8	9
650	81 291	298	305	311	318	325	331	338	345	351
651	358	365	371	378	385	391	398	405	411	418
652	425	431	438	445	451	458	465	471	478	485
653	491	498	505	511	518	525	531	538	544	551
654	558	564	571	578	584	591	598	604	611	617
655	624	631	637	644	651	657	664	671	677	684
656	690	697	704	710	717	723	730	737	743	750
657	757	763	770	776	783	790	796	803	809	816
658	823	829	836	842	849	856	862	869	875	882
659	889	895	902	908	915	921	928	935	941	948
660	954	961	968	974	981	987	994	*000	*007	*014
661	82 020	027	033	040	046	053	060	066	073	079
662	086	092	099	105	112	119	125	132	138	145
663	151	158	164	171	178	184	191	197	204	210
664	217	223	230	236	243	249	256	263	269	276
665	282	289	295	302	308	315	321	328	334	341
666	347	354	360	367	373	380	387	393	400	406
667	413	419	426	432	439	445	452	458	465	471
668	478	484	491	497	504	510	517	523	530	536
669	543	549	556	562	569	575	582	588	595	601
670	607	614	620	627	633	640	646	653	659	666
671	672	679	685	692	698	705	711	718	724	730
672	737	743	750	756	763	769	776	782	789	795
673	802	808	814	821	827	834	840	847	853	860
674	866	872	879	885	892	898	905	911	918	924
675	930	937	943	950	956	963	969	975	982	988
676	995	*001	*008	*014	*020	*027	*033	*040	*046	*052
677	83 059	065	072	078	085	091	097	104	110	117
678	123	129	136	142	149	155	161	168	174	181
679	187	193	200	206	213	219	225	232	238	245
680	251	257	264	270	276	283	289	296	302	308
681	315	321	327	334	340	347	353	359	366	372
682	378	385	391	398	404	410	417	423	429	436
683	442	448	455	461	467	474	480	487	493	499
684	506	512	518	525	531	537	544	550	556	563
685	569	575	582	588	594	601	607	613	620	626
686	632	639	645	651	658	664	670	677	683	689
687	696	702	708	715	721	727	734	740	746	753
688	759	765	771	778	784	790	797	803	809	816
689	822	828	835	841	847	853	860	866	872	879
690	885	891	897	904	910	916	923	929	935	942
691	948	954	960	967	973	979	985	992	998	*004
692	84 011	017	023	029	036	042	048	055	061	067
693	073	080	086	092	098	105	111	117	123	130
694	136	142	148	155	161	167	173	180	186	192
695	198	205	211	217	223	230	236	242	248	255
696	261	267	273	280	286	292	298	305	311	317
697	323	330	336	342	348	354	361	367	373	379
698	386	392	398	404	410	417	423	429	435	442
699	448	454	460	466	473	479	485	491	497	504
700	510	516	522	528	535	541	547	553	559	566
N.	L. 0	1	2	3	4	5	6	7	8	9

Prop. Parts

	7
1	0.7
2	1.4
3	2.1
4	2.8
5	3.5
6	4.2
7	4.9
8	5.6
9	6.3

	6
1	0.6
2	1.2
3	1.8
4	2.4
5	3.0
6	3.6
7	4.2
8	4.8
9	5.4

N.	L. 0	1	2	3	4	5	6	7	8	9	Prop. Parts
700	84 510	516	522	528	535	541	547	553	559	566	
701	572	578	584	590	597	603	609	615	621	628	
702	634	640	646	652	658	665	671	677	683	689	
703	696	702	708	714	720	726	733	739	745	751	
704	757	763	770	776	782	788	794	800	807	813	
705	819	825	831	837	844	850	856	862	868	874	
706	880	887	893	899	905	911	917	924	930	936	7
707	942	948	954	960	967	973	979	985	991	997	1 0.7
708	85 003	009	016	022	028	034	040	046	052	058	2 1.4
709	065	071	077	083	089	095	101	107	114	120	3 2.1
710	126	132	138	144	150	156	163	169	175	181	4 2.8
711	187	193	199	205	211	217	224	230	236	242	5 3.5
712	248	254	260	266	272	278	285	291	297	303	6 4.2
713	309	315	321	327	333	339	345	352	358	364	7 4.9
714	370	376	382	388	394	400	406	412	418	425	8 5.6
715	431	437	443	449	455	461	467	473	479	485	9 6.3
716	491	497	503	509	516	522	528	534	540	546	
717	552	558	564	570	576	582	588	594	600	606	
718	612	618	625	631	637	643	649	655	661	667	
719	673	679	685	691	697	703	709	715	721	727	
720	733	739	745	751	757	763	769	775	781	788	
721	794	800	806	812	818	824	830	836	842	848	6
722	854	860	866	872	878	884	890	896	902	908	1 0.6
723	914	920	926	932	938	944	950	956	962	968	2 1.2
724	974	980	986	992	998	*004	*010	*016	*022	*028	3 1.8
725	86 034	040	046	052	058	064	070	076	082	088	4 2.4
726	094	100	106	112	118	124	130	136	141	147	5 3.0
727	153	159	165	171	177	183	189	195	201	207	6 3.6
728	213	219	225	231	237	243	249	255	261	267	7 4.2
729	273	279	285	291	297	303	308	314	320	326	8 4.8
730	332	338	344	350	356	362	368	374	380	386	9 5.4
731	392	398	404	410	415	421	427	433	439	445	
732	451	457	463	469	475	481	487	493	499	504	
733	510	516	522	528	534	540	546	552	558	564	
734	570	576	581	587	593	599	605	611	617	623	
735	629	635	641	646	652	658	664	670	676	682	
736	688	694	700	705	711	717	723	729	735	741	5
737	747	753	759	764	770	776	782	788	794	800	1 0.5
738	806	812	817	823	829	835	841	847	853	859	2 1.0
739	864	870	876	882	888	894	900	906	911	917	3 1.5
740	923	929	935	941	947	953	958	964	970	976	4 2.0
741	982	988	994	999	*005	*011	*017	*023	*029	*035	5 2.5
742	87 040	046	052	058	064	070	075	081	087	093	6 3.0
743	099	105	111	116	122	128	134	140	146	151	7 3.5
744	157	163	169	175	181	186	192	198	204	210	8 4.0
745	216	221	227	233	239	245	251	256	262	268	9 4.5
746	274	280	286	291	297	303	309	315	320	326	
747	332	338	344	349	355	361	367	373	379	384	
748	390	396	402	408	413	419	425	431	437	442	
749	448	454	460	466	471	477	483	489	495	500	
750	506	512	518	523	529	535	541	547	552	558	
N.	L. 0	1	2	3	4	5	6	7	8	9	Prop. Parts

750-800

N.	L. 0	1	2	3	4	5	6	7	8	9
750	87 506	512	518	523	529	535	541	547	552	558
751	564	570	576	581	587	593	599	604	610	616
752	622	628	633	639	64$\bar{5}$	651	656	662	668	674
753	679	685	691	697	703	708	714	720	726	731
754	737	743	749	754	760	766	772	777	783	789
755	79$\bar{5}$	800	806	812	818	823	829	83$\bar{5}$	841	846
756	852	858	864	869	875	881	887	892	898	904
757	910	915	921	927	933	938	944	9$\bar{5}$0	955	961
758	967	973	978	984	990	996	*001	*007	*013	*018
759	88 024	030	036	041	047	053	058	064	070	076
760	081	087	093	098	104	110	116	121	127	133
761	138	144	1$\bar{5}$0	156	161	167	173	178	184	190
762	195	201	207	213	218	224	230	235	241	247
763	252	258	264	270	275	281	287	292	298	304
764	309	315	321	326	332	338	343	349	35$\bar{5}$	360
765	366	372	377	383	389	39$\bar{5}$	400	406	412	417
766	423	429	434	440	446	451	457	463	468	474
767	480	485	491	497	502	508	513	519	52$\bar{5}$	530
768	536	542	547	553	559	564	570	576	581	587
769	593	598	604	610	615	621	627	632	638	643
770	649	65$\bar{5}$	660	666	672	677	683	689	694	700
771	705	711	717	722	728	734	739	74$\bar{5}$	750	756
772	762	767	773	779	784	790	795	801	807	812
773	818	824	829	83$\bar{5}$	840	846	852	857	863	868
774	874	880	885	891	897	902	908	913	919	92$\bar{5}$
775	930	936	941	947	953	958	964	969	97$\bar{5}$	981
776	986	992	997	*003	*009	*014	*020	*025	*031	*037
777	89 042	048	053	059	064	070	076	081	087	092
778	098	104	109	11$\bar{5}$	120	126	131	137	143	148
779	154	159	16$\bar{5}$	170	176	182	187	193	198	204
780	209	215	221	226	232	237	243	248	254	260
781	265	271	276	282	287	293	298	304	310	315
782	321	326	332	337	343	348	354	360	365	371
783	376	382	387	393	398	404	409	41$\bar{5}$	421	426
784	432	437	443	448	454	459	46$\bar{5}$	470	476	481
785	487	492	498	504	509	51$\bar{5}$	520	526	531	537
786	542	548	553	559	564	570	575	581	586	592
787	597	603	609	614	620	625	631	636	642	647
788	653	658	664	669	67$\bar{5}$	680	686	691	697	702
789	708	713	719	724	730	735	741	746	752	757
790	763	768	774	779	78$\bar{5}$	790	796	801	807	812
791	818	823	829	834	840	845	851	856	862	867
792	873	878	883	889	894	900	905	911	916	922
793	927	933	938	944	949	95$\bar{5}$	960	966	971	977
794	982	988	993	998	*004	*009	*01$\bar{5}$	*020	*026	*031
795	90 037	042	048	053	059	064	069	07$\bar{5}$	080	086
796	091	097	102	108	113	119	124	129	13$\bar{5}$	140
797	146	151	157	162	168	173	179	184	189	19$\bar{5}$
798	200	206	211	217	222	227	233	238	244	249
799	25$\bar{5}$	260	266	271	276	282	287	293	298	304
800	309	314	320	325	331	336	342	347	352	358
N.	L. 0	1	2	3	4	5	6	7	8	9

Prop. Parts

	6
1	0.6
2	1.2
3	1.8
4	2.4
5	3.0
6	3.6
7	4.2
8	4.8
9	5.4

	5
1	0.5
2	1.0
3	1.5
4	2.0
5	2.5
6	3.0
7	3.5
8	4.0
9	4.5

N.	L. 0	1	2	3	4	5	6	7	8	9	Prop. Parts
800	90 309	314	320	325	331	336	342	347	352	358	
801	363	369	374	380	385	390	396	401	407	412	
802	417	423	428	434	439	445	450	455	461	466	
803	472	477	482	488	493	499	504	509	515	520	
804	526	531	536	542	547	553	558	563	569	574	
805	580	585	590	596	601	607	612	617	623	628	
806	634	639	644	650	655	660	666	671	677	682	
807	687	693	698	703	709	714	720	725	730	736	
808	741	747	752	757	763	768	773	779	784	789	
809	795	800	806	811	816	822	827	832	838	843	
810	849	854	859	865	870	875	881	886	891	897	
811	902	907	913	918	924	929	934	940	945	950	**6**
812	956	961	966	972	977	982	988	993	998	*004	1 | 0.6
813	91 009	014	020	025	030	036	041	046	052	057	2 | 1.2
814	062	068	073	078	084	089	094	100	105	110	3 | 1.8
815	116	121	126	132	137	142	148	153	158	164	4 | 2.4
816	169	174	180	185	190	196	201	206	212	217	5 | 3.0
817	222	228	233	238	243	249	254	259	265	270	6 | 3.6
818	275	281	286	291	297	302	307	312	318	323	7 | 4.2
819	328	334	339	344	350	355	360	365	371	376	8 | 4.8
820	381	387	392	397	403	408	413	418	424	429	9 | 5.4
821	434	440	445	450	455	461	466	471	477	482	
822	487	492	498	503	508	514	519	524	529	535	
823	540	545	551	556	561	566	572	577	582	587	
824	593	598	603	609	614	619	624	630	635	640	
825	645	651	656	661	666	672	677	682	687	693	
826	698	703	709	714	719	724	730	735	740	745	
827	751	756	761	766	772	777	782	787	793	798	
828	803	808	814	819	824	829	834	840	845	850	
829	855	861	866	871	876	882	887	892	897	903	
830	908	913	918	924	929	934	939	944	950	955	
831	960	965	971	976	981	986	991	997	*002	*007	**5**
832	92 012	018	023	028	033	038	044	049	054	059	1 | 0.5
833	065	070	075	080	085	091	096	101	106	111	2 | 1.0
834	117	122	127	132	137	143	148	153	158	163	3 | 1.5
835	169	174	179	184	189	195	200	205	210	215	4 | 2.0
836	221	226	231	236	241	247	252	257	262	267	5 | 2.5
837	273	278	283	288	293	298	304	309	314	319	6 | 3.0
838	324	330	335	340	345	350	355	361	366	371	7 | 3.5
839	376	381	387	392	397	402	407	412	418	423	8 | 4.0
840	428	433	438	443	449	454	459	464	469	474	9 | 4.5
841	480	485	490	495	500	505	511	516	521	526	
842	531	536	542	547	552	557	562	567	572	578	
843	583	588	593	598	603	609	614	619	624	629	
844	634	639	645	650	655	660	665	670	675	681	
845	686	691	696	701	706	711	716	722	727	732	
846	737	742	747	752	758	763	768	773	778	783	
847	788	793	799	804	809	814	819	824	829	834	
848	840	845	850	855	860	865	870	875	881	886	
849	891	896	901	906	911	916	921	927	932	937	
850	942	947	952	957	962	967	973	978	983	988	
N.	L. 0	1	2	3	4	5	6	7	8	9	Prop. Parts

N.	L. 0	1	2	3	4	5	6	7	8	9
850	92 942	947	952	957	962	967	973	978	983	988
851	993	998	*003	*008	*013	*018	*024	*029	*034	*039
852	93 044	049	054	059	064	069	075̄	080	085̄	090
853	095̄	100	105	110	115	120	125	131	136	141
854	146	151	156	161	166	171	176	181	186	192
855	197	202	207	212	217	222	227	232	237	242
856	247	252	258	263	268	273	278	283	288	293
857	298	303	308	313	318	323	328	334	339	344
858	349	354	359	364	369	374	379	384	389	394
859	399	404	409	414	420	425̄	430	435̄	440	445̄
860	45̄0	455̄	460	465̄	470	475	480	485	490	495
861	500	505	510	515	520	526	531	536	541	546
862	551	556	561	566	571	576	581	586	591	596
863	601	606	611	616	621	626	631	636	641	646
864	651	656	661	666	671	676	682	687	692	697
865	702	707	712	717	722	727	732	737	742	747
866	752	757	762	767	772	777	782	787	792	797
867	802	807	812	817	822	827	832	837	842	847
868	852	857	862	867	872	877	882	887	892	897
869	902	907	912	917	922	927	932	937	942	947
870	952	957	962	967	972	977	982	987	992	997
871	94 002	007	012	017	022	027	032	037	042	047
872	052	057	062	067	072	077	082	086	091	096
873	101	106	111	116	121	126	131	136	141	146
874	151	156	161	166	171	176	181	186	191	196
875	201	206	211	216	221	226	231	236	240	245
876	250	255	260	265	270	275	280	285	290	295
877	300	305̄	310	315̄	320	325̄	330	335̄	340	345̄
878	349	354	359	364	369	374	379	384	389	394
879	399	404	409	414	419	424	429	433	438	443
880	448	453	458	463	468	473	478	483	488	493
881	498	503	507	512	517	522	527	532	537	542
882	547	552	557	562	567	571	576	581	586	591
883	596	601	606	611	616	621	626	630	635	640
884	645	650	655	660	665̄	670	675̄	680	685̄	689
885	694	699	704	709	714	719	724	729	734	738
886	743	748	753	758	763	768	773	778	783	787
887	792	797	802	807	812	817	822	827	832	836
888	841	846	851	856	861	866	871	876	880	885
889	890	895	900	905̄	910	915̄	919	924	929	934
890	939	944	949	954	959	963	968	973	978	983
891	988	993	998	*002	*007	*012	*017	*022	*027	*032
892	95 036	041	046	051	056	061	066	071	075	080
893	085	090	095̄	100	105̄	109	114	119	124	129
894	134	139	143	148	153	158	163	168	173	177
895	182	187	192	197	202	207	211	216	221	226
896	231	236	240	245	250	255	260	265̄	270	274
897	279	284	289	294	299	303	308	313	318	323
898	328	332	337	342	347	352	357	361	366	371
899	376	381	386	390	395	400	405̄	410	415̄	419
900	424	429	434	439	444	448	453	458	463	468
N.	L. 0	1	2	3	4	5	6	7	8	9

Prop. Parts

	6
1	0.6
2	1.2
3	1.8
4	2.4
5	3.0
6	3.6
7	4.2
8	4.8
9	5.4

	5
1	0.5
2	1.0
3	1.5
4	2.0
5	2.5
6	3.0
7	3.5
8	4.0
9	4.5

	4
1	0.4
2	0.8
3	1.2
4	1.6
5	2.0
6	2.4
7	2.8
8	3.2
9	3.6

900-950

N.	L. 0	1	2	3	4	5	6	7	8	9	Prop. Parts
900	95 424	429	434	439	444	448	453	458	463	468	
901	472	477	482	487	492	497	501	506	511	516	
902	521	525	530	535	540	54$\bar{5}$	5$\bar{5}$0	554	559	564	
903	569	574	578	583	588	593	598	602	607	612	
904	617	622	626	631	636	641	646	650	655	660	
905	66$\bar{5}$	670	674	679	684	689	694	698	703	708	
906	713	718	722	727	732	737	742	746	751	756	
907	761	766	770	775	780	78$\bar{5}$	789	794	799	804	
908	809	813	818	823	828	832	837	842	847	852	
909	856	861	866	871	875	880	885	890	89$\bar{5}$	899	
910	904	909	914	918	923	928	933	938	942	947	
911	952	957	961	966	971	976	980	985	990	99$\bar{5}$	**5**
912	999	*004	*009	*014	*019	*023	*028	*033	*038	*042	1 | 0.5
913	96 047	052	057	061	066	071	076	080	085	090	2 | 1.0
914	09$\bar{5}$	099	104	109	114	118	123	128	133	137	3 | 1.5
915	142	147	152	156	161	166	171	175	180	18$\bar{5}$	4 | 2.0
916	190	194	199	204	209	213	218	223	227	232	5 | 2.5
917	237	242	246	251	256	261	265	270	27$\bar{5}$	280	6 | 3.0
918	284	289	294	298	303	308	313	317	322	327	7 | 3.5
919	332	336	341	346	350	355	360	36$\bar{5}$	369	374	8 | 4.0
920	379	384	388	393	398	402	407	412	417	421	9 | 4.5
921	426	431	435	440	44$\bar{5}$	4$\bar{5}$0	454	459	464	468	
922	473	478	483	487	492	497	501	506	511	515	
923	520	52$\bar{5}$	530	534	539	544	548	553	558	562	
924	567	572	577	581	586	591	595	600	60$\bar{5}$	609	
925	614	619	624	628	633	638	642	647	652	656	
926	661	666	670	675	680	68$\bar{5}$	689	694	699	703	
927	708	713	717	722	727	731	736	741	745	750	
928	75$\bar{5}$	759	764	769	774	778	783	788	792	797	
929	802	806	811	816	820	82$\bar{5}$	830	834	839	844	
930	848	853	858	862	867	872	876	881	886	890	
931	89$\bar{5}$	900	904	909	914	918	923	928	932	937	**4**
932	942	946	951	956	960	96$\bar{5}$	970	974	979	984	1 | 0.4
933	988	993	997	*002	*007	*011	*016	*021	*025	*030	2 | 0.8
934	97 03$\bar{5}$	039	044	049	053	058	063	067	072	077	3 | 1.2
935	081	086	090	095	100	104	109	114	118	123	4 | 1.6
936	128	132	137	142	146	151	155	160	16$\bar{5}$	169	5 | 2.0
937	174	179	183	188	192	197	202	206	211	216	6 | 2.4
938	220	22$\bar{5}$	230	234	239	243	248	253	257	262	7 | 2.8
939	267	271	276	280	285	290	294	299	304	308	8 | 3.2
940	313	317	322	327	331	336	340	345	3$\bar{5}$0	354	9 | 3.6
941	359	364	368	373	377	382	387	391	396	400	
942	405	410	414	419	424	428	433	437	442	447	
943	451	456	460	46$\bar{5}$	470	474	479	483	488	493	
944	497	502	506	511	516	520	52$\bar{5}$	529	534	539	
945	543	548	552	557	562	566	571	575	580	58$\bar{5}$	
946	589	594	598	603	607	612	617	621	626	630	
947	63$\bar{5}$	640	644	649	653	658	663	667	672	676	
948	681	685	690	69$\bar{5}$	699	704	708	713	717	722	
949	727	731	736	740	74$\bar{5}$	749	754	759	763	768	
950	772	777	782	786	791	795	800	804	809	813	
N.	L. 0	1	2	3	4	5	6	7	8	9	Prop. Parts

950-1000

N.	L. 0	1	2	3	4	5	6	7	8	9
950	97 772	777	782	786	791	795	800	804	809	813
951	818	823	827	832	836	841	845	850	85$\bar{5}$	859
952	864	868	873	877	882	886	891	896	900	90$\bar{5}$
953	909	914	918	923	928	932	937	941	946	950
954	95$\bar{5}$	959	964	968	973	978	982	987	991	996
955	98 000	00$\bar{5}$	009	014	019	023	028	032	037	041
956	046	050	05$\bar{5}$	059	064	068	073	078	082	087
957	091	096	100	10$\bar{5}$	109	114	118	123	127	132
958	137	141	146	150	15$\bar{5}$	159	164	168	173	177
959	182	186	191	195	200	204	209	214	218	223
960	227	232	236	241	245	2$\bar{5}$0	254	259	263	268
961	272	277	281	286	290	29$\bar{5}$	299	304	308	313
962	318	322	327	331	336	340	34$\bar{5}$	349	354	358
963	363	367	372	376	381	385	390	394	399	403
964	408	412	417	421	426	430	43$\bar{5}$	439	444	448
965	453	457	462	466	471	475	480	484	489	493
966	498	502	507	511	516	520	52$\bar{5}$	529	534	538
967	543	547	552	556	561	565	570	574	579	583
968	588	592	597	601	605	610	614	619	623	628
969	632	637	641	646	650	65$\bar{5}$	659	664	668	673
970	677	682	686	691	695	700	704	709	713	717
971	722	726	731	735	740	744	749	753	758	762
972	767	771	776	780	784	789	793	798	802	807
973	811	816	820	82$\bar{5}$	829	834	838	843	847	851
974	856	860	86$\bar{5}$	869	874	878	883	887	892	896
975	900	90$\bar{5}$	909	914	918	923	927	932	936	941
976	94$\bar{5}$	949	954	958	963	967	972	976	981	985
977	989	994	998	*003	*007	*012	*016	*021	*025	*029
978	99 034	038	043	047	052	056	061	06$\bar{5}$	069	074
979	078	083	087	092	096	100	10$\bar{5}$	109	114	118
980	123	127	131	136	140	14$\bar{5}$	149	154	158	162
981	167	171	176	180	18$\bar{5}$	189	193	198	202	207
982	211	216	220	224	229	233	238	242	247	251
983	255	260	264	269	273	277	282	286	291	295
984	300	304	308	313	317	322	326	330	33$\bar{5}$	339
985	344	348	352	357	361	366	370	374	379	383
986	388	392	396	401	405	410	414	419	423	427
987	432	436	441	44$\bar{5}$	449	454	458	463	467	471
988	476	480	484	489	493	498	502	506	511	515
989	520	524	528	533	537	542	546	550	55$\bar{5}$	559
990	564	568	572	577	581	585	590	594	599	603
991	607	612	616	621	62$\bar{5}$	629	634	638	642	647
992	651	656	660	664	669	673	677	682	686	691
993	69$\bar{5}$	699	704	708	712	717	721	726	730	734
994	739	743	747	752	756	760	76$\bar{5}$	769	774	778
995	782	787	791	795	800	804	808	813	817	822
996	826	830	83$\bar{5}$	839	843	848	852	856	861	865
997	870	874	878	883	887	891	896	900	904	909
998	913	917	922	926	930	93$\bar{5}$	939	944	948	952
999	957	961	965	970	974	978	983	987	991	996
1000	00 000	004	009	013	017	022	026	030	03$\bar{5}$	039
N.	L. 0	1	2	3	4	5	6	7	8	9

Prop. Parts

	5
1	0.5
2	1.0
3	1.5
4	2.0
5	2.5
6	3.0
7	3.5
8	4.0
9	4.5

	4
1	0.4
2	0.8
3	1.2
4	1.6
5	2.0
6	2.4
7	2.8
8	3.2
9	3.6

TABLE II

LOGARITHMS OF TRIGONOMETRIC FUNCTIONS

″	′	*l* sin	log *S*	*l* csc	*l* tan	log *T*	*l* cot	*l* sec	*l* cos	′
0	**0**	Inf. neg.	—	Infinite.	Inf. neg.	—	Infinite.	10.00000	10.00000	**60**
60	1	6.46373	5.31 443	13.53627	6.46373	5.31 443	13.53627	00000	00000	59
120	2	76476	5.31 443	23524	76476	5.31 443	23524	00000	00000	58
180	3	94085	5.31 443	05915	94085	5.31 443	05915	00000	00000	57
240	4	7.06579	5.31 443	12.93421	7.06579	5.31 442	12.93421	00000	00000	56
300	**5**	7.16270	5.31 443	12.83730	7.16270	5.31 442	12.83730	10.00000	10.00000	**55**
360	6	24188	5.31 443	75812	24188	5.31 442	75812	00000	00000	54
420	7	30882	5.31 443	69118	30882	5.31 442	69118	00000	00000	53
480	8	36682	5.31 443	63318	36682	5.31 442	63318	00000	00000	52
540	9	41797	5.31 443	58203	41797	5.31 442	58203	00000	00000	51
600	**10**	7.46373	5.31 443	12.53627	7.46373	5.31 442	12.53627	10.00000	10.00000	**50**
660	11	50512	5.31 443	49488	50512	5.31 442	49488	00000	00000	49
720	12	54291	5.31 443	45709	54291	5.31 442	45709	00000	00000	48
780	13	57767	5.31 443	42233	57767	5.31 442	42233	00000	00000	47
840	14	60985	5.31 443	39015	60986	5.31 442	39014	00000	00000	46
900	**15**	7.63982	5.31 443	12.36018	7.63982	5.31 442	12.36018	10.00000	10.00000	**45**
960	16	66784	5.31 443	33216	66785	5.31 442	33215	00000	00000	44
1020	17	69417	5.31 443	30583	69418	5.31 442	30582	00001	9.99999	43
1080	18	71900	5.31 443	28100	71900	5.31 442	28100	00001	99999	42
1140	19	74248	5.31 443	25752	74248	5.31 442	25752	00001	99999	41
1200	**20**	7.76475	5.31 443	12.23525	7.76476	5.31 442	12.23524	10.00001	9.99999	**40**
1260	21	78594	5.31 443	21406	78595	5.31 442	21405	00001	99999	39
1320	22	80615	5.31 443	19385	80615	5.31 442	19385	00001	99999	38
1380	23	82545	5.31 443	17455	82546	5.31 442	17454	00001	99999	37
1440	24	84393	5.31 443	15607	84394	5.31 442	15606	00001	99999	36
1500	**25**	7.86166	5.31 443	12.13834	7.86167	5.31 442	12.13833	10.00001	9.99999	**35**
1560	26	87870	5.31 443	12130	87871	5.31 442	12129	00001	99999	34
1620	27	89509	5.31 443	10491	89510	5.31 442	10490	00001	99999	33
1680	28	91088	5.31 443	08912	91089	5.31 442	08911	00001	99999	32
1740	29	92612	5.31 443	07388	92613	5.31 441	07387	00002	99998	31
1800	**30**	7.94084	5.31 443	12.05916	7.94086	5.31 441	12.05914	10.00002	9.99998	**30**
1860	31	95508	5.31 443	04492	95510	5.31 441	04490	00002	99998	29
1920	32	96887	5.31 443	03113	96889	5.31 441	03111	00002	99998	28
1980	33	98223	5.31 443	01777	98225	5.31 441	01775	00002	99998	27
2040	34	99520	5.31 443	00480	99522	5.31 441	00478	00002	99998	26
2100	**35**	8.00779	5.31 443	11.99221	8.00781	5.31 441	11.99219	10.00002	9.99998	**25**
2160	36	02002	5.31 443	97998	02004	5.31 441	97996	00002	99998	24
2220	37	03192	5.31 443	96808	03194	5.31 441	96806	00003	99997	23
2280	38	04350	5.31 443	95650	04353	5.31 441	95647	00003	99997	22
2340	39	05478	5.31 443	94522	05481	5.31 441	94519	00003	99997	21
2400	**40**	8.06578	5.31 443	11.93422	8.06581	5.31 441	11.93419	10.00003	9.99997	**20**
2460	41	07650	5.31 444	92350	07653	5.31 440	92347	00003	99997	19
2520	42	08696	5.31 444	91304	08700	5.31 440	91300	00003	99997	18
2580	43	09718	5.31 444	90282	09722	5.31 440	90278	00003	99997	17
2640	44	10717	5.31 444	89283	10720	5.31 440	89280	00004	99996	16
2700	**45**	8.11693	5.31 444	11.88307	8.11696	5.31 440	11.88304	10.00004	9.99996	**15**
2760	46	12647	5.31 444	87353	12651	5.31 440	87349	00004	99996	14
2820	47	13581	5.31 444	86419	13585	5.31 440	86415	00004	99996	13
2880	48	14495	5.31 444	85505	14500	5.31 440	85500	00004	99996	12
2940	49	15391	5.31 444	84609	15395	5.31 440	84605	00004	99996	11
3000	**50**	8.16268	5.31 444	11.83732	8.16273	5.31 439	11.83727	10.00005	9.99995	**10**
3060	51	17128	5.31 444	82872	17133	5.31 439	82867	00005	99995	9
3120	52	17971	5.31 444	82029	17976	5.31 439	82024	00005	99995	8
3180	53	18798	5.31 444	81202	18804	5.31 439	81196	00005	99995	7
3240	54	19610	5.31 444	80390	19616	5.31 439	80384	00005	99995	6
3300	**55**	8.20407	5.31 444	11.79593	8.20413	5.31 439	11.79587	10.00006	9.99994	**5**
3360	56	21189	5.31 444	78811	21195	5.31 439	78805	00006	99994	4
3420	57	21958	5.31 445	78042	21964	5.31 439	78036	00006	99994	3
3480	58	22713	5.31 445	77287	22720	5.31 438	77280	00006	99994	2
3540	59	23456	5.31 445	76544	23462	5.31 438	76538	00006	99994	1
3600	**60**	24186	5.31 445	75814	24192	5.31 438	75808	00007	99993	**0**
	′	*l* cos		*l* sec	*l* cot		*l* tan	*l* csc	*l* sin	′

″	′	l sin	log S	l csc	l tan	log T	l cot	l sec	l cos	′
3600	**0**	8.24186	5.31 445	11.75814	8.24192	5.31 438	11.75808	10.00007	9.99993	**60**
3660	1	24903	5.31 445	75097	24910	5.31 438	75090	00007	99993	59
3720	2	25609	5.31 445	74391	25616	5.31 438	74384	00007	99993	58
3780	3	26304	5.31 445	73696	26312	5.31 438	73688	00007	99993	57
3840	4	26988	5.31 445	73012	26996	5.31 437	73004	00008	99992	56
3900	**5**	8.27661	5.31 445	11.72339	8.27669	5.31 437	11.72331	10.00008	9.99992	**55**
3960	6	28324	5.31 445	71676	28332	5.31 437	71668	00008	99992	54
4020	7	28977	5.31 445	71023	28986	5.31 437	71014	00008	99992	53
4080	8	29621	5.31 445	70379	29629	5.31 437	70371	00008	99992	52
4140	9	30255	5.31 445	69745	30263	5.31 437	69737	00009	99991	51
4200	**10**	8.30879	5.31 446	11.69121	8.30888	5.31 437	11.69112	10.00009	9.99991	**50**
4260	11	31495	5.31 446	68505	31505	5.31 436	68495	00009	99991	49
4320	12	32103	5.31 446	67897	32112	5.31 436	67888	00010	99990	48
4380	13	32702	5.31 446	67298	32711	5.31 436	67289	00010	99990	47
4440	14	33292	5.31 446	66708	33302	5.31 436	66698	00010	99990	46
4500	**15**	8.33875	5.31 446	11.66125	8.33886	5.31 436	11.66114	10.00010	9.99990	**45**
4560	16	34450	5.31 446	65550	34461	5.31 435	65539	00011	99989	44
4620	17	35018	5.31 446	64982	35029	5.31 435	64971	00011	99989	43
4680	18	35578	5.31 446	64422	35590	5.31 435	64410	00011	99989	42
4740	19	36131	5.31 446	63869	36143	5.31 435	63857	00011	99989	41
4800	**20**	8.36678	5.31 446	11.63322	8.36689	5.31 435	11.63311	10.00012	9.99988	**40**
4860	21	37217	5.31 447	62783	37229	5.31 434	62771	00012	99988	39
4920	22	37750	5.31 447	62250	37762	5.31 434	62238	00012	99988	38
4980	23	38276	5.31 447	61724	38289	5.31 434	61711	00013	99987	37
5040	24	38796	5.31 447	61204	38809	5.31 434	61191	00013	99987	36
5100	**25**	8.39310	5.31 447	11.60690	8.39323	5.31 434	11.60677	10.00013	9.99987	**35**
5160	26	39818	5.31 447	60182	39832	5.31 433	60168	00014	99986	34
5220	27	40320	5.31 447	59680	40334	5.31 433	59666	00014	99986	33
5280	28	40816	5.31 447	59184	40830	5.31 433	59170	00014	99986	32
5340	29	41307	5.31 447	58693	41321	5.31 433	58679	00015	99985	31
5400	**30**	8.41792	5.31 447	11.58208	8.41807	5.31 433	11.58193	10.00015	9.99985	**30**
5460	31	42272	5.31 448	57728	42287	5.31 432	57713	00015	99985	29
5520	32	42746	5.31 448	57254	42762	5.31 432	57238	00016	99984	28
5580	33	43216	5.31 448	56784	43232	5.31 432	56768	00016	99984	27
5640	34	43680	5.31 448	56320	43696	5.31 432	56304	00016	99984	26
5700	**35**	8.44139	5.31 448	11.55861	8.44156	5.31 431	11.55844	10.00017	9.99983	**25**
5760	36	44594	5.31 448	55406	44611	5.31 431	55389	00017	99983	24
5820	37	45044	5.31 448	54956	45061	5.31 431	54939	00017	99983	23
5880	38	45489	5.31 448	54511	45507	5.31 431	54493	00018	99982	22
5940	39	45930	5.31 449	54070	45948	5.31 431	54052	00018	99982	21
6000	**40**	8.46366	5.31 449	11.53634	8.46385	5.31 430	11.53615	10.00018	9.99982	**20**
6060	41	46799	5.31 449	53201	46817	5.31 430	53183	00019	99981	19
6120	42	47226	5.31 449	52774	47245	5.31 430	52755	00019	99981	18
6180	43	47650	5.31 449	52350	47669	5.31 430	52331	00019	99981	17
6240	44	48069	5.31 449	51931	48089	5.31 429	51911	00020	99980	16
6300	**45**	8.48485	5.31 449	11.51515	8.48505	5.31 429	11.51495	10.00020	9.99980	**15**
6360	46	48896	5.31 449	51104	48917	5.31 429	51083	00021	99979	14
6420	47	49304	5.31 450	50696	49325	5.31 428	50675	00021	99979	13
6480	48	49708	5.31 450	50292	49729	5.31 428	50271	00021	99979	12
6540	49	50108	5.31 450	49892	50130	5.31 428	49870	00022	99978	11
6600	**50**	8.50504	5.31 450	11.49496	8.50527	5.31 428	11.49473	10.00022	9.99978	**10**
6660	51	50897	5.31 450	49103	50920	5.31 427	49080	00023	99977	9
6720	52	51287	5.31 450	48713	51310	5.31 427	48690	00023	99977	8
6780	53	51673	5.31 450	48327	51696	5.31 427	48304	00023	99977	7
6840	54	52055	5.31 450	47945	52079	5.31 427	47921	00024	99976	6
6900	**55**	8.52434	5.31 451	11.47566	8.52459	5.31 426	11.47541	10.00024	9.99976	**5**
6960	56	52810	5.31 451	47190	52835	5.31 426	47165	00025	99975	4
7020	57	53183	5.31 451	46817	53208	5.31 426	46792	00025	99975	3
7080	58	53552	5.31 451	46448	53578	5.31 425	46422	00026	99974	2
7140	59	53919	5.31 451	46081	53945	5.31 425	46055	00026	99974	1
7200	**60**	54282	5.31 451	45718	54308	5.31 425	45692	00026	99974	**0**
	′	l cos		l sec	l cot		l tan	l csc	l sin	′

″	′	l sin	log S	l csc	l tan	log T	l cot	l sec	d 1′	l cos	′
7200	**0**	8.54282	5.31 451	11.45718	8.54308	5.31 425	11.45692	10.00026	1	9.99974	**60**
7260	1	54642	5.31 451	45358	54669	5.31 425	45331	00027	0	99973	59
7320	2	54999	5.31 452	45001	55027	5.31 424	44973	00027	1	99973	58
7380	3	55354	5.31 452	44646	55382	5.31 424	44618	00028	0	99972	57
7440	4	55705	5.31 452	44295	55734	5.31 424	44266	00028	1	99972	56
7500	**5**	8.56054	5.31 452	11.43946	8.56083	5.31 423	11.43917	10.00029	0	9.99971	**55**
7560	6	56400	5.31 452	43600	56429	5.31 423	43571	00029	1	99971	54
7620	7	56743	5.31 452	43257	56773	5.31 423	43227	00030	0	99970	53
7680	8	57084	5.31 453	42916	57114	5.31 422	42886	00030	1	99970	52
7740	9	57421	5.31 453	42579	57452	5.31 422	42548	00031	0	99969	51
7800	**10**	8.57757	5.31 453	11.42243	8.57788	5.31 422	11.42212	10.00031	1	9.99969	**50**
7860	11	58089	5.31 453	41911	58121	5.31 421	41879	00032	0	99968	49
7920	12	58419	5.31 453	41581	58451	5.31 421	41549	00032	1	99968	48
7980	13	58747	5.31 453	41253	58779	5.31 421	41221	00033	0	99967	47
8040	14	59072	5.31 454	40928	59105	5.31 421	40895	00033	0	99967	46
8100	**15**	8.59395	5.31 454	11.40605	8.59428	5.31 420	11.40572	10.00033	1	9.99967	**45**
8160	16	59715	5.31 454	40285	59749	5.31 420	40251	00034	0	99966	44
8220	17	60033	5.31 454	39967	60068	5.31 420	39932	00034	1	99966	43
8280	18	60349	5.31 454	39651	60384	5.31 419	39616	00035	1	99965	42
8340	19	60662	5.31 454	39338	60698	5.31 419	39302	00036	0	99964	41
8400	**20**	8.60973	5.31 455	11.39027	8.61009	5.31 418	11.38991	10.00036	1	9.99964	**40**
8460	21	61282	5.31 455	38718	61319	5.31 418	38681	00037	0	99963	39
8520	22	61589	5.31 455	38411	61626	5.31 418	38374	00037	1	99963	38
8580	23	61894	5.31 455	38106	61931	5.31 417	38069	00038	0	99962	37
8640	24	62196	5.31 455	37804	62234	5.31 417	37766	00038	1	99962	36
8700	**25**	8.62497	5.31 455	11.37503	8.62535	5.31 417	11.37465	10.00039	0	9.99961	**35**
8760	26	62795	5.31 456	37205	62834	5.31 416	37166	00039	1	99961	34
8820	27	63091	5.31 456	36909	63131	5.31 416	36869	00040	0	99960	33
8880	28	63385	5.31 456	36615	63426	5.31 416	36574	00040	1	99960	32
8940	29	63678	5.31 456	36322	63718	5.31 415	36282	00041	0	99959	31
9000	**30**	8.63968	5.31 456	11.36032	8.64009	5.31 415	11.35991	10.00041	1	9.99959	**30**
9060	31	64256	5.31 456	35744	64298	5.31 415	35702	00042	0	99958	29
9120	32	64543	5.31 457	35457	64585	5.31 414	35415	00042	1	99958	28
9180	33	64827	5.31 457	35173	64870	5.31 414	35130	00043	1	99957	27
9240	34	65110	5.31 457	34890	65154	5.31 413	34846	00044	0	99956	26
9300	**35**	8.65391	5.31 457	11.34609	8.65435	5.31 413	11.34565	10.00044	1	9.99956	**25**
9360	36	65670	5.31 457	34330	65715	5.31 413	34285	00045	0	99955	24
9420	37	65947	5.31 458	34053	65993	5.31 412	34007	00045	1	99955	23
9480	38	66223	5.31 458	33777	66269	5.31 412	33731	00046	0	99954	22
9540	39	66497	5.31 458	33503	66543	5.31 412	33457	00046	1	99954	21
9600	**40**	8.66769	5.31 458	11.33231	8.66816	5.31 411	11.33184	10.00047	1	9.99953	**20**
9660	41	67039	5.31 458	32961	67087	5.31 411	32913	00048	0	99952	19
9720	42	67308	5.31 459	32692	67356	5.31 410	32644	00048	1	99952	18
9780	43	67575	5.31 459	32425	67624	5.31 410	32376	00049	0	99951	17
9840	44	67841	5.31 459	32159	67890	5.31 410	32110	00049	1	99951	16
9900	**45**	8.68104	5.31 459	11.31896	8.68154	5.31 409	11.31846	10.00050	1	9.99950	**15**
9960	46	68367	5.31 459	31633	68417	5.31 409	31583	00051	0	99949	14
10020	47	68627	5.31 460	31373	68678	5.31 408	31322	00051	1	99949	13
10080	48	68886	5.31 460	31114	68938	5.31 408	31062	00052	0	99948	12
10140	49	69144	5.31 460	30856	69196	5.31 408	30804	00052	1	99948	11
10200	**50**	8.69400	5.31 460	11.30600	8.69453	5.31 407	11.30547	10.00053	1	9.99947	**10**
10260	51	69654	5.31 460	30346	69708	5.31 407	30292	00054	0	99946	9
10320	52	69907	5.31 461	30093	69962	5.31 406	30038	00054	1	99946	8
10380	53	70159	5.31 461	29841	70214	5.31 406	29786	00055	1	99945	7
10440	54	70409	5.31 461	29591	70465	5.31 405	29535	00056	0	99944	6
10500	**55**	8.70658	5.31 461	11.29342	8.70714	5.31 405	11.29286	10.00056	1	9.99944	**5**
10560	56	70905	5.31 461	29095	70962	5.31 405	29038	00057	1	99943	4
10620	57	71151	5.31 462	28849	71208	5.31 404	28792	00058	0	99942	3
10680	58	71395	5.31 462	28605	71453	5.31 404	28547	00058	1	99942	2
10740	59	71638	5.31 462	28362	71697	5.31 403	28303	00059	1	99941	1
10800	**60**	71880	5.31 462	28120	71940	5.31 403	28060	00060		99940	**0**
	′	l cos		l sec	l cot		l tan	l csc	d 1′	l sin	′

′	l sin 8.	d 1′	l csc 11.	l tan 8.	d 1′	l cot 11.	l sec 10.	d 1′	l cos 9.	′
0	71880	240	28120	71940	241	28060	00060	0	99940	60
1	72120	239	27880	72181	239	27819	060	1	940	59
2	359	238	641	420	239	580	061	1	939	58
3	597	237	403	659	237	341	062	0	938	57
4	834	235	166	896	236	104	062	1	938	56
5	73069	234	26931	73132	234	26868	063	1	937	55
6	303	232	697	366	234	634	064	0	936	54
7	535	232	465	600	232	400	064	1	936	53
8	767	230	233	832	231	168	065	1	935	52
9	997	229	003	74063	229	25937	066	0	934	51
10	74226	228	25774	292	229	708	066	1	934	50
11	454	226	546	521	227	479	067	1	933	49
12	680	226	320	748	226	252	068	0	932	48
13	906	224	094	974	225	026	068	1	932	47
14	75130	223	24870	75199	224	24801	069	1	931	46
15	353	222	647	423	222	577	070	1	930	45
16	575	220	425	645	222	355	071	0	929	44
17	795	220	205	867	220	133	071	1	929	43
18	76015	219	23985	76087	219	23913	072	1	928	42
19	234	217	766	306	219	694	073	1	927	41
20	451	216	549	525	217	475	074	0	926	40
21	667	216	333	742	216	258	074	1	926	39
22	883	214	117	958	215	042	075	1	925	38
23	77097	213	22903	77173	214	22827	076	1	924	37
24	310	212	690	387	213	613	077	0	923	36
25	522	211	478	600	211	400	077	1	923	35
26	733	210	267	811	211	189	078	1	922	34
27	943	209	057	78022	210	21978	079	1	921	33
28	78152	208	21848	232	209	768	080	0	920	32
29	360	208	640	441	208	559	080	1	920	31
30	78568	206	21432	78649	206	21351	00081	1	99919	30
31	774	205	226	855	206	145	082	1	918	29
32	979	204	021	79061	205	20939	083	0	917	28
33	79183	203	20817	266	204	734	083	1	917	27
34	386	202	614	470	203	530	084	1	916	26
35	588	201	412	673	202	327	085	1	915	25
36	789	201	211	875	201	125	086	1	914	24
37	990	199	010	80076	201	19924	087	0	913	23
38	80189	199	19811	277	199	723	087	1	913	22
39	388	197	612	476	198	524	088	1	912	21
40	585	197	415	674	198	326	089	1	911	20
41	782	196	218	872	196	128	090	1	910	19
42	978	195	022	81068	196	18932	091	0	909	18
43	81173	194	18827	264	195	736	091	1	909	17
44	367	193	633	459	194	541	092	1	908	16
45	560	192	440	653	193	347	093	1	907	15
46	752	192	248	846	192	154	094	1	906	14
47	944	190	056	82038	192	17962	095	1	905	13
48	82134	190	17866	230	190	770	096	0	904	12
49	324	189	676	420	190	580	096	1	904	11
50	513	188	487	610	189	390	097	1	903	10
51	701	187	299	799	188	201	098	1	902	9
52	888	187	112	987	188	013	099	1	901	8
53	83075	186	16925	83175	186	16825	100	1	900	7
54	261	185	739	361	186	639	101	1	899	6
55	446	184	554	547	185	453	102	0	898	5
56	630	183	370	732	184	268	102	1	898	4
57	813	183	187	916	184	084	103	1	897	3
58	996	181	004	84100	182	15900	104	1	896	2
59	84177	181	15823	282	182	718	105	1	895	1
60	84358		15642	84464		15536	00106		99894	0
′	8. l cos	d 1′	11. l sec	8. l cot	d 1′	11. l tan	10. l csc	d 1′	9. l sin	′

″	Proportional Parts 241	239	237	235	234	232	229
0	0	0	0	0	0	0	0
1	4	4	4	4	4	4	4
2	8	8	8	8	8	8	8
3	12	12	12	12	12	12	11
4	16	16	16	16	16	15	15
5	20	20	20	20	19	19	19
6	24	24	24	24	23	23	23
7	28	28	28	27	27	27	27
8	32	32	32	31	31	31	31
9	36	36	36	35	35	35	34
10	40	40	40	39	39	39	38
11	44	44	43	43	43	43	42
12	48	48	47	47	47	46	46
13	52	52	51	51	51	50	50
14	56	56	55	55	55	54	53
15	60	60	59	59	59	58	57
16	64	64	63	63	62	62	61
17	68	68	67	67	66	66	65
18	72	72	71	70	70	70	69
19	76	76	75	74	74	73	73
20	80	80	79	78	78	77	76
21	84	84	83	82	82	81	80
22	88	88	87	86	86	85	84
23	92	92	91	90	90	89	88
24	96	96	95	94	94	93	92
25	100	100	99	98	97	97	95
26	104	104	103	102	101	101	99
27	108	108	107	106	105	104	103
28	112	112	111	110	109	108	107
29	116	116	115	114	113	112	111
30	120	120	118	118	117	116	114
31	125	123	122	121	121	120	118
32	129	127	126	125	125	124	122
33	133	131	130	129	129	128	126
34	137	135	134	133	133	131	130
35	141	139	138	137	137	135	134
36	145	143	142	141	140	139	137
37	149	147	146	145	144	143	141
38	153	151	150	149	148	147	145
39	157	155	154	153	152	151	149
40	161	159	158	157	156	155	153
41	165	163	162	161	160	159	156
42	169	167	166	164	164	162	160
43	173	171	170	168	168	166	164
44	177	175	174	172	172	170	168
45	181	179	178	176	175	174	172
46	185	183	182	180	179	178	176
47	189	187	186	184	183	182	179
48	193	191	190	188	187	186	183
49	197	195	194	192	191	189	187
50	201	199	198	196	195	193	191
51	205	203	201	200	199	197	195
52	209	207	205	204	203	201	198
53	213	211	209	208	207	205	202
54	217	215	213	212	211	209	206
55	221	219	217	215	215	213	210
56	225	223	221	219	218	217	214
57	229	227	225	223	222	220	218
58	233	231	229	227	226	224	221
59	237	235	233	231	230	228	225
60	241	239	237	235	234	232	229
″	241 Proportional Parts	239	237	235	234	232	229

93° 86°

TABLE II

″	Proportional Parts																					
	227	**225**	**223**	**220**	**217**	**215**	**213**	**211**	**208**	**206**	**203**	**201**	**199**	**197**	**195**	**193**	**192**	**189**	**187**	**185**	**183**	**181**
0	0	0	0	0	0	0	0	0	0	0	0	0	0	0	0	0	0	0	0	0	0	0
1	4	4	4	4	4	4	4	4	3	3	3	3	3	3	3	3	3	3	3	3	3	3
2	8	8	7	7	7	7	7	7	7	7	7	7	7	7	6	6	6	6	6	6	6	6
3	11	11	11	11	11	11	11	11	10	10	10	10	10	10	10	10	10	9	9	9	9	9
4	15	15	15	15	14	14	14	14	14	14	14	13	13	13	13	13	13	13	12	12	12	12
5	19	19	19	18	18	18	18	18	17	17	17	17	17	16	16	16	16	16	16	15	15	15
6	23	22	22	22	22	22	21	21	21	21	20	20	20	20	20	19	19	19	19	18	18	18
7	26	26	26	26	25	25	25	25	24	24	24	23	23	23	23	23	22	22	22	22	21	21
8	30	30	30	29	29	29	28	28	28	27	27	27	27	26	26	26	26	25	25	25	24	24
9	34	34	33	33	33	32	32	32	31	31	30	30	30	30	29	29	29	28	28	28	27	27
10	38	38	37	37	36	36	36	35	35	34	34	34	33	33	32	32	32	32	31	31	30	30
11	42	41	41	40	40	39	39	39	38	38	37	37	36	36	36	35	35	35	34	34	34	33
12	45	45	45	44	43	43	43	42	42	41	41	40	40	39	39	39	38	38	37	37	37	36
13	49	49	48	48	47	47	46	46	45	45	44	44	43	43	42	42	42	41	41	40	40	39
14	53	52	52	51	51	50	50	49	49	48	47	47	46	46	46	45	45	44	44	43	43	42
15	57	56	56	55	54	54	53	53	52	51	51	50	50	49	49	48	48	47	47	46	46	45
16	61	60	59	59	58	57	57	56	55	55	54	54	53	53	52	51	51	50	50	49	49	48
17	64	64	63	62	61	61	60	60	59	58	58	57	56	56	55	55	54	54	53	52	52	51
18	68	68	67	66	65	64	64	63	62	62	61	60	60	59	58	58	58	57	56	56	55	54
19	72	71	71	70	69	68	67	67	66	65	64	64	63	62	62	61	61	60	59	59	58	57
20	76	75	74	73	72	72	71	70	69	69	68	67	66	66	65	64	64	63	62	62	61	60
21	79	79	78	77	76	75	75	74	73	72	71	70	70	69	68	68	67	66	65	65	64	63
22	83	82	82	81	80	79	78	77	76	76	74	74	73	72	72	71	70	69	69	68	67	66
23	87	86	85	84	83	82	82	81	80	79	78	77	76	76	75	74	74	72	72	71	70	69
24	91	90	89	88	87	86	85	84	83	82	81	80	80	79	78	77	77	76	75	74	73	72
25	95	94	93	92	90	90	89	88	87	86	85	84	83	82	81	80	80	79	78	77	76	75
26	98	98	97	95	94	93	92	91	90	89	88	87	86	85	84	84	83	82	81	80	79	78
27	102	101	100	99	98	97	96	95	94	93	91	90	90	89	88	87	86	85	84	83	82	81
28	106	105	104	103	101	100	99	98	97	96	95	94	93	92	91	90	90	88	87	86	85	84
29	110	109	108	106	105	104	103	102	101	100	98	97	96	95	94	93	93	91	90	89	88	87
30	114	112	112	110	108	108	106	106	104	103	102	100	100	98	98	96	96	94	94	92	92	90
31	117	116	115	114	112	111	110	109	107	106	105	104	103	102	101	100	99	98	97	96	95	94
32	121	120	119	117	116	115	114	113	111	110	108	107	106	105	104	103	102	101	100	99	98	97
33	125	124	123	121	119	118	117	116	114	113	112	111	109	108	107	106	106	104	103	102	101	100
34	129	128	126	125	123	122	121	120	118	117	115	114	113	112	110	109	109	107	106	105	104	103
35	132	131	130	128	127	125	124	123	121	120	118	117	116	115	114	113	112	110	109	108	107	106
36	136	135	134	132	130	129	128	127	125	124	122	121	119	118	117	116	115	113	112	111	110	109
37	140	139	138	136	134	133	131	130	128	127	125	124	123	121	120	119	118	117	115	114	113	112
38	144	142	141	139	137	136	135	134	132	130	129	127	126	125	124	122	122	120	118	117	116	115
39	148	146	145	143	141	140	138	137	135	134	132	131	129	128	127	125	125	123	122	120	119	118
40	151	150	149	147	145	143	142	141	139	137	135	134	133	131	130	129	128	126	125	123	122	121
41	155	154	152	150	148	147	146	144	142	141	139	137	136	135	133	132	131	129	128	126	125	124
42	159	158	156	154	152	150	149	148	146	144	142	141	139	138	136	135	134	132	131	130	128	127
43	163	161	160	158	156	154	153	151	149	148	145	144	143	141	140	138	138	135	134	133	131	130
44	166	165	164	161	159	158	156	155	153	151	149	147	146	144	143	142	141	139	137	136	134	133
45	170	169	167	165	163	161	160	158	156	155	152	151	149	148	146	145	144	142	140	139	137	136
46	174	172	171	169	166	165	163	162	159	158	156	154	153	151	150	148	147	145	143	142	140	139
47	178	176	175	172	170	168	167	165	163	161	159	157	156	154	153	151	150	148	146	145	143	142
48	182	180	178	176	174	172	170	169	166	165	162	161	159	158	156	154	154	151	150	148	146	145
49	185	184	182	180	177	176	174	172	170	168	166	164	163	161	159	158	157	154	153	151	149	148
50	189	188	186	183	181	179	178	176	173	172	169	168	166	164	162	161	160	158	156	154	152	151
51	193	191	190	187	184	183	181	179	177	175	173	171	169	167	166	164	163	161	159	157	156	154
52	197	195	193	191	188	186	185	183	180	179	176	174	172	171	169	167	166	164	162	160	159	157
53	201	199	197	194	192	190	188	186	184	182	179	178	176	174	172	170	170	167	165	163	162	160
54	204	202	201	198	195	194	192	190	187	185	183	181	179	177	176	174	173	170	168	166	165	163
55	208	206	204	202	199	197	195	193	191	189	186	184	182	181	179	177	176	173	171	170	168	166
56	212	210	208	205	203	201	199	197	194	192	189	188	186	184	182	180	179	176	175	173	171	169
57	216	214	212	209	206	204	202	200	198	196	193	191	189	187	185	183	182	180	178	176	174	172
58	219	218	216	213	210	208	206	204	201	199	196	194	192	190	188	187	186	183	181	179	177	175
59	223	221	219	216	213	211	209	207	205	203	200	198	196	194	192	190	189	186	184	182	180	178
60	227	225	223	220	217	215	213	211	208	206	203	201	199	197	195	193	192	189	187	185	183	181
″	**227**	**225**	**223**	**220**	**217**	**215**	**213**	**211**	**208**	**206**	**203**	**201**	**199**	**197**	**195**	**193**	**192**	**189**	**187**	**185**	**183**	**181**
	Proportional Parts																					

′	l sin 8.	d 1′	l csc 11.	l tan 8.	d 1′	l cot 11.	l sec 10.	d 1′	l cos 9.	′
0	84358	181	15642	84464	182	15536	00106	1	99894	60
1	539	179	461	646	180	354	107	1	893	59
2	718	179	282	826	180	174	108	1	892	58
3	897	178	103	85006	179	14994	109	0	891	57
4	85075	177	14925	185	178	815	109	1	891	56
5	252	177	748	363	177	637	110	1	890	55
6	429	176	571	540	177	460	111	1	889	54
7	605	175	395	717	176	283	112	1	888	53
8	780	175	220	893	176	107	113	1	887	52
9	955	173	045	86069	174	13931	114	1	886	51
10	86128	173	13872	243	174	757	115	1	885	50
11	301	173	699	417	174	583	116	1	884	49
12	474	171	526	591	172	409	117	1	883	48
13	645	171	355	763	172	237	118	1	882	47
14	816	171	184	935	171	065	119	1	881	46
15	987	169	013	87106	171	12894	120	1	880	45
16	87156	169	12844	277	170	723	121	0	879	44
17	325	169	675	447	169	553	121	1	879	43
18	494	167	506	616	169	384	122	1	878	42
19	661	168	339	785	168	215	123	1	877	41
20	829	166	171	953	167	047	124	1	876	40
21	995	166	005	88120	167	11880	125	1	875	39
22	88161	165	11839	287	166	713	126	1	874	38
23	326	164	674	453	165	547	127	1	873	37
24	490	164	510	618	165	382	128	1	872	36
25	654	163	346	783	165	217	129	1	871	35
26	817	163	183	948	163	052	130	1	870	34
27	980	162	020	89111	163	10889	131	1	869	33
28	89142	162	10858	274	163	726	132	1	868	32
29	304	160	696	437	161	563	133	1	867	31
30	89464	161	10536	89598	162	10402	00134	1	99866	30
31	625	159	375	760	160	240	135	1	865	29
32	784	159	216	920	160	080	136	1	864	28
33	943	159	057	90080	160	09920	137	1	863	27
34	90102	158	09898	240	159	760	138	1	862	26
35	260	157	740	399	158	601	139	1	861	25
36	417	157	583	557	158	443	140	1	860	24
37	574	156	426	715	157	285	141	1	859	23
38	730	155	270	872	157	128	142	1	858	22
39	885	155	115	91029	156	08971	143	1	857	21
40	91040	155	08960	185	155	815	144	1	856	20
41	195	154	805	340	155	660	145	1	855	19
42	349	153	651	495	155	505	146	1	854	18
43	502	153	498	650	153	350	147	1	853	17
44	655	152	345	803	154	197	148	1	852	16
45	807	152	193	957	153	043	149	1	851	15
46	959	151	041	92110	152	07890	150	2	850	14
47	92110	151	07890	262	152	738	152	1	848	13
48	261	150	739	414	151	586	153	1	847	12
49	411	150	589	565	151	435	154	1	846	11
50	561	149	439	716	150	284	155	1	845	10
51	710	149	290	866	150	134	156	1	844	9
52	859	148	141	93016	149	06984	157	1	843	8
53	93007	147	06993	165	148	835	158	1	842	7
54	154	147	846	313	149	687	159	1	841	6
55	301	147	699	462	147	538	160	1	840	5
56	448	146	552	609	147	391	161	1	839	4
57	594	146	406	756	147	244	162	1	838	3
58	740	145	260	903	146	097	163	1	837	2
59	885	145	115	94049	146	05951	164	2	836	1
60	94030		05970	94195		05805	00166		99834	0
′	8. l cos	d 1′	11. l sec	8. l cot	d 1′	11. l tan	10. l csc	d 1′	9. l sin	′

″	182	181	179	177	176	175	174
0	0	0	0	0	0	0	0
1	3	3	3	3	3	3	3
2	6	6	6	6	6	6	6
3	9	9	9	9	9	9	9
4	12	12	12	12	12	12	12
5	15	15	15	15	15	15	14
6	18	18	18	18	18	18	17
7	21	21	21	21	21	20	20
8	24	24	24	24	23	23	23
9	27	27	27	27	26	26	26
10	30	30	30	30	29	29	29
11	33	33	33	32	32	32	32
12	36	36	36	35	35	35	35
13	39	39	39	38	38	38	38
14	42	42	42	41	41	41	41
15	45	45	45	44	44	44	44
16	49	48	48	47	47	47	46
17	52	51	51	50	50	50	49
18	55	54	54	53	53	52	52
19	58	57	57	56	56	55	55
20	61	60	60	59	59	58	58
21	64	63	63	62	62	61	61
22	67	66	66	65	65	64	64
23	70	69	69	68	67	67	67
24	73	72	72	71	70	70	70
25	76	75	75	74	73	73	72
26	79	78	78	77	76	76	75
27	82	81	81	80	79	79	78
28	85	84	84	83	82	82	81
29	88	87	87	86	85	85	84
30	91	90	90	88	88	88	87
31	94	94	92	91	91	90	90
32	97	97	95	94	94	93	93
33	100	100	98	97	97	96	96
34	103	103	101	100	100	99	99
35	106	106	104	103	103	102	102
36	109	109	107	106	106	105	104
37	112	112	110	109	109	108	107
38	115	115	113	112	111	111	110
39	118	118	116	115	114	114	113
40	121	121	119	118	117	117	116
41	124	124	122	121	120	120	119
42	127	127	125	124	123	122	122
43	130	130	128	127	126	125	125
44	133	133	131	130	129	128	128
45	137	136	134	133	132	131	130
46	140	139	137	136	135	134	133
47	143	142	140	139	138	137	136
48	146	145	143	142	141	140	139
49	149	148	146	145	144	143	142
50	152	151	149	148	147	146	145
51	155	154	152	150	150	149	148
52	158	157	155	153	153	152	151
53	161	160	158	156	155	155	154
54	164	163	161	159	158	158	157
55	167	166	164	162	161	160	160
56	170	169	167	165	164	163	162
57	173	172	170	168	167	166	165
58	176	175	173	171	170	169	168
59	179	178	176	174	173	172	171
60	182	181	179	177	176	175	174
″	182	181	179	177	176	175	174

Proportional Parts

94° 85°

TABLE II

"	Proportional Parts																					
	173	**172**	**171**	**169**	**167**	**166**	**165**	**163**	**162**	**160**	**159**	**158**	**157**	**155**	**153**	**152**	**151**	**150**	**149**	**147**	**146**	**145**
0	0	0	0	0	0	0	0	0	0	0	0	0	0	0	0	0	0	0	0	0	0	0
1	3	3	3	3	3	3	3	3	3	3	3	3	3	3	3	3	3	2	2	2	2	2
2	6	6	6	6	6	6	6	5	5	5	5	5	5	5	5	5	5	5	5	5	5	5
3	9	9	9	8	8	8	8	8	8	8	8	8	8	8	8	8	8	8	7	7	7	7
4	12	11	11	11	11	11	11	11	11	11	11	11	10	10	10	10	10	10	10	10	10	10
5	14	14	14	14	14	14	14	14	14	13	13	13	13	13	13	13	13	12	12	12	12	12
6	17	17	17	17	17	17	16	16	16	16	16	16	16	16	15	15	15	15	15	15	15	14
7	20	20	20	20	19	19	19	19	19	19	19	18	18	18	18	18	18	18	17	17	17	17
8	23	23	23	23	22	22	22	22	22	21	21	21	21	21	20	20	20	20	20	20	19	19
9	26	26	26	25	25	25	25	24	24	24	24	24	24	23	23	23	23	22	22	22	22	22
10	29	29	28	28	28	28	28	27	27	27	26	26	26	26	26	25	25	25	25	24	24	24
11	32	32	31	31	31	30	30	30	30	29	29	29	29	28	28	28	28	28	27	27	27	27
12	35	34	34	34	33	33	33	33	32	32	32	32	31	31	31	30	30	30	30	29	29	29
13	37	37	37	37	36	36	36	35	35	35	34	34	34	34	33	33	33	32	32	32	32	31
14	40	40	40	39	39	39	38	38	38	37	37	37	37	36	36	35	35	35	35	34	34	34
15	43	43	43	42	42	42	41	41	40	40	40	40	39	39	38	38	38	38	37	37	36	36
16	46	46	46	45	45	44	44	43	43	43	42	42	42	41	41	41	40	40	40	39	39	39
17	49	49	48	48	47	47	47	46	46	45	45	45	44	44	43	43	43	42	42	42	41	41
18	52	52	51	51	50	50	50	49	49	48	48	47	47	46	46	46	45	45	45	44	44	44
19	55	54	54	54	53	53	52	52	51	51	50	50	50	49	48	48	48	48	47	47	46	46
20	58	57	57	56	56	55	55	54	54	53	53	53	52	52	51	51	50	50	50	49	49	48
21	61	60	60	59	58	58	58	57	57	56	56	55	55	54	54	53	53	52	52	51	51	51
22	63	63	63	62	61	61	60	60	59	59	58	58	58	57	56	56	55	55	55	54	54	53
23	66	66	66	65	64	64	63	62	62	61	61	61	60	59	59	58	58	58	57	56	56	56
24	69	69	68	68	67	66	66	65	65	64	64	63	63	62	61	61	60	60	60	59	58	58
25	72	72	71	70	70	69	69	68	68	67	66	66	65	65	64	63	63	62	62	61	61	60
26	75	75	74	73	72	72	72	71	70	69	69	68	68	67	66	66	65	65	65	64	63	63
27	78	77	77	76	75	75	74	73	73	72	72	71	71	70	69	68	68	68	67	66	66	65
28	81	80	80	79	78	77	77	76	76	75	74	74	73	72	71	71	70	70	70	69	68	68
29	84	83	83	82	81	80	80	79	78	77	77	76	76	75	74	73	73	72	72	71	71	70
30	86	86	86	84	84	83	82	82	81	80	80	79	78	78	76	76	76	75	74	74	73	72
31	89	89	88	87	86	86	85	84	84	83	82	82	81	80	79	79	78	78	77	76	75	75
32	92	92	91	90	89	89	88	87	86	85	85	84	84	83	82	81	81	80	79	78	78	77
33	95	95	94	93	92	91	91	90	89	88	87	87	86	85	84	84	83	82	82	81	80	80
34	98	97	97	96	95	94	94	92	92	91	90	90	89	88	87	86	86	85	84	83	83	82
35	101	100	100	99	97	97	96	95	94	93	93	92	92	90	89	89	88	88	87	86	85	85
36	104	103	103	101	100	100	99	98	97	96	95	95	94	93	92	91	91	90	89	88	88	87
37	107	106	105	104	103	102	102	101	100	99	98	97	97	96	94	94	93	92	92	91	90	89
38	110	109	108	107	106	105	104	103	103	101	101	100	99	98	97	96	96	95	94	93	92	92
39	112	112	111	110	109	108	107	106	105	104	103	103	102	101	99	99	98	98	97	96	95	94
40	115	115	114	113	111	111	110	109	108	107	106	105	105	103	102	101	101	100	99	98	97	97
41	118	118	117	115	114	113	113	111	111	109	109	108	107	106	105	104	103	102	102	100	100	99
42	121	120	120	118	117	116	116	114	113	112	111	111	110	108	107	106	106	105	104	103	102	102
43	124	123	123	121	120	119	118	117	116	115	114	113	113	111	110	109	108	108	107	105	105	104
44	127	126	125	124	122	122	121	120	119	117	117	116	115	114	112	111	111	110	109	108	107	106
45	130	129	128	127	125	124	124	122	122	120	119	118	118	116	115	114	113	112	112	110	110	109
46	133	132	131	130	128	127	126	125	124	123	122	121	120	119	117	117	116	115	114	113	112	111
47	136	135	134	132	131	130	129	128	127	125	125	124	123	121	120	119	118	118	117	115	114	114
48	138	138	137	135	134	133	132	130	130	128	127	126	126	124	122	122	121	120	119	118	117	116
49	141	140	140	138	136	136	135	133	132	131	130	129	128	127	125	124	123	122	122	120	119	118
50	144	143	142	141	139	138	138	136	135	133	132	132	131	129	128	127	126	125	124	122	122	121
51	147	146	145	144	142	141	140	139	138	136	135	134	133	132	130	129	128	128	127	125	124	123
52	150	149	148	146	145	144	143	141	140	139	138	137	136	134	133	132	131	130	129	127	127	126
53	153	152	151	149	148	147	146	144	143	141	140	140	139	137	135	134	133	132	132	130	129	128
54	156	155	154	152	150	149	148	147	146	144	143	142	141	140	138	137	136	135	134	132	131	130
55	159	158	157	155	153	152	151	149	148	147	146	145	144	142	140	139	138	138	137	135	134	133
56	161	161	160	158	156	155	154	152	151	149	148	147	147	145	143	142	141	140	139	137	136	135
57	164	163	162	161	159	158	157	155	154	152	151	150	149	147	145	144	143	142	142	140	139	138
58	167	166	165	163	161	160	160	158	157	155	154	153	152	150	148	147	146	145	144	142	141	140
59	170	169	168	166	164	163	162	160	159	157	156	155	154	152	150	149	148	148	147	145	144	143
60	173	172	171	169	167	166	165	163	162	160	159	158	157	155	153	152	151	150	149	147	146	145
"	**173**	**172**	**171**	**169**	**167**	**166**	**165**	**163**	**162**	**160**	**159**	**158**	**157**	**155**	**153**	**152**	**151**	**150**	**149**	**147**	**146**	**145**
	Proportional Parts																					

′	l sin 8.	d 1′	l csc 11.	l tan 8.	d 1′	l cot 11.	l sec 10.	d 1′	l cos 9.	′
0	94030	144	05970	94195	145	05805	00166	1	99834	60
1	174	143	826	340	145	660	167	1	833	59
2	317	144	683	485	145	515	168	1	832	58
3	461	142	539	630	143	370	169	1	831	57
4	603	143	397	773	144	227	170	1	830	56
5	746	141	254	917	143	083	171	1	829	55
6	887	142	113	95060	142	04940	172	1	828	54
7	95029	141	04971	202	142	798	173	2	827	53
8	170	140	830	344	142	656	175	1	825	52
9	310	140	690	486	141	514	176	1	824	51
10	450	139	550	627	140	373	177	1	823	50
11	589	139	411	767	141	233	178	1	822	49
12	728	139	272	908	139	092	179	1	821	48
13	867	138	133	96047	140	03953	180	1	820	47
14	96005	138	03995	187	138	813	181	2	819	46
15	143	137	857	325	139	675	183	1	817	45
16	280	137	720	464	138	536	184	1	816	44
17	417	136	583	602	137	398	185	1	815	43
18	553	136	447	739	138	261	186	1	814	42
19	689	136	311	877	136	123	187	1	813	41
20	825	135	175	97013	137	02987	188	2	812	40
21	960	135	040	150	135	850	190	1	810	39
22	97095	134	02905	285	136	715	191	1	809	38
23	229	134	771	421	135	579	192	1	808	37
24	363	133	637	556	135	444	193	1	807	36
25	496	133	504	691	134	309	194	2	806	35
26	629	133	371	825	134	175	196	1	804	34
27	762	132	238	959	133	041	197	1	803	33
28	894	132	106	98092	133	01908	198	1	802	32
29	98026	131	01974	225	133	775	199	1	801	31
30	98157	131	01843	98358	132	01642	00200	2	99800	30
31	288	131	712	490	132	510	202	1	798	29
32	419	130	581	622	131	378	203	1	797	28
33	549	130	451	753	131	247	204	1	796	27
34	679	129	321	884	131	116	205	2	795	26
35	808	129	192	99015	130	00985	207	1	793	25
36	937	129	063	145	130	855	208	1	792	24
37	99066	128	00934	275	130	725	209	1	791	23
38	194	128	806	405	129	595	210	2	790	22
39	322	128	678	534	128	466	212	1	788	21
40	450	127	550	662	129	338	213	1	787	20
41	577	127	423	791	128	209	214	1	786	19
42	704	126	296	919	127	081	215	2	785	18
43	830	126	170	00046	128	99954	217	1	783	17
44	956	126	044	174	127	826	218	1	782	16
45	00082	125	99918	301	126	699	219	1	781	15
46	207	125	793	427	126	573	220	2	780	14
47	332	124	668	553	126	447	222	1	778	13
48	456	125	544	679	126	321	223	1	777	12
49	581	123	419	805	125	195	224	1	776	11
50	704	124	296	930	125	070	225	2	775	10
51	828	123	172	01055	124	98945	227	1	773	9
52	951	123	049	179	124	821	228	1	772	8
53	01074	122	98926	303	124	697	229	2	771	7
54	196	122	804	427	123	573	231	1	769	6
55	318	122	682	550	123	450	232	1	768	5
56	440	121	560	673	123	327	233	2	767	4
57	561	121	439	796	122	204	235	1	765	3
58	682	121	318	918	122	082	236	1	764	2
59	803	120	197	02040	122	97960	237	2	763	1
60	01923		98077	02162		97838	00239		99761	0
′	9. l cos	d 1′	10. l sec	9. l cot	d 1′	10. l tan	10. l csc	d 1′	9. l sin	′

″	Proportional Parts						
	145	144	143	142	141	140	139
0	0	0	0	0	0	0	0
1	2	2	2	2	2	2	2
2	5	5	5	5	5	5	5
3	7	7	7	7	7	7	7
4	10	10	10	9	9	9	9
5	12	12	12	12	12	12	12
6	14	14	14	14	14	14	14
7	17	17	17	17	16	16	16
8	19	19	19	19	19	19	19
9	22	22	21	21	21	21	21
10	24	24	24	24	24	23	23
11	27	26	26	26	26	26	25
12	29	29	29	28	28	28	28
13	31	31	31	31	31	30	30
14	34	34	33	33	33	33	32
15	36	36	36	36	35	35	35
16	39	38	38	38	38	37	37
17	41	41	41	40	40	40	39
18	44	43	43	43	42	42	42
19	46	46	45	45	45	44	44
20	48	48	48	47	47	47	46
21	51	50	50	50	49	49	49
22	53	53	52	52	52	51	51
23	56	55	55	54	54	54	53
24	58	58	57	57	56	56	56
25	60	60	60	59	59	58	58
26	63	62	62	62	61	61	60
27	65	65	64	64	63	63	63
28	68	67	67	66	66	65	65
29	70	70	69	69	68	68	67
30	72	72	72	71	70	70	70
31	75	74	74	73	73	72	72
32	77	77	76	76	75	75	74
33	80	79	79	78	78	77	76
34	82	82	81	80	80	79	79
35	85	84	83	83	82	82	81
36	87	86	86	85	85	84	83
37	89	89	88	88	87	86	86
38	92	91	91	90	89	89	88
39	94	94	93	92	92	91	90
40	97	96	95	95	94	93	93
41	99	98	98	97	96	96	95
42	102	101	100	99	99	98	97
43	104	103	102	102	101	100	100
44	106	106	105	104	103	103	102
45	109	108	107	106	106	105	104
46	111	110	110	109	108	107	107
47	114	113	112	111	110	110	109
48	116	115	114	114	113	112	111
49	118	118	117	116	115	114	114
50	121	120	119	118	118	117	116
51	123	122	122	121	120	119	118
52	126	125	124	123	122	121	120
53	128	127	126	125	125	124	123
54	130	130	129	128	127	126	125
55	133	132	131	130	129	128	127
56	135	134	133	133	132	131	130
57	138	137	136	135	134	133	132
58	140	139	138	137	136	135	134
59	143	142	141	140	139	138	137
60	145	144	143	142	141	140	139
″	145	144	143	142	141	140	139
	Proportional Parts						

TABLE II

″	Proportional Parts																			
″	**138**	**137**	**136**	**135**	**134**	**133**	**132**	**131**	**130**	**129**	**128**	**127**	**126**	**125**	**124**	**123**	**122**	**121**	**120**	**2**	**1**
0	0	0	0	0	0	0	0	0	0	0	0	0	0	0	0	0	0	0	0	0	0
1	2	2	2	2	2	2	2	2	2	2	2	2	2	2	2	2	2	2	2	0	0
2	5	5	5	4	4	4	4	4	4	4	4	4	4	4	4	4	4	4	4	0	0
3	7	7	7	7	7	7	7	7	6	6	6	6	6	6	6	6	6	6	6	0	0
4	9	9	9	9	9	9	9	9	9	9	9	8	8	8	8	8	8	8	8	0	0
5	12	11	11	11	11	11	11	11	11	11	11	11	10	10	10	10	10	10	10	0	0
6	14	14	14	14	13	13	13	13	13	13	13	13	13	12	12	12	12	12	12	0	0
7	16	16	16	16	16	16	15	15	15	15	15	15	15	15	14	14	14	14	14	0	0
8	18	18	18	18	18	18	18	17	17	17	17	17	17	17	17	16	16	16	16	0	0
9	21	21	20	20	20	20	20	20	20	19	19	19	19	19	19	18	18	18	18	0	0
10	23	23	23	22	22	22	22	22	22	22	21	21	21	21	21	20	20	20	20	0	0
11	25	25	25	25	25	24	24	24	24	24	23	23	23	23	23	23	22	22	22	0	0
12	28	27	27	27	27	27	26	26	26	26	26	25	25	25	25	25	24	24	24	0	0
13	30	30	29	29	29	29	29	28	28	28	28	28	27	27	27	27	26	26	26	0	0
14	32	32	32	32	31	31	31	31	30	30	30	30	29	29	29	29	28	28	28	0	0
15	34	34	34	34	34	33	33	33	32	32	32	32	32	31	31	31	30	30	30	0	0
16	37	37	36	36	36	35	35	35	35	34	34	34	34	33	33	33	33	32	32	1	0
17	39	39	39	38	38	38	37	37	37	37	36	36	36	35	35	35	35	34	34	1	0
18	41	41	41	40	40	40	40	39	39	39	38	38	38	38	37	37	37	36	36	1	0
19	44	43	43	43	42	42	42	41	41	41	41	40	40	40	39	39	39	38	38	1	0
20	46	46	45	45	45	44	44	44	43	43	43	42	42	42	41	41	41	40	40	1	0
21	48	48	48	47	47	47	46	46	46	45	45	44	44	44	43	43	43	42	42	1	0
22	51	50	50	50	49	49	48	48	48	47	47	47	46	46	45	45	45	44	44	1	0
23	53	53	52	52	51	51	51	50	50	49	49	49	48	48	48	47	47	46	46	1	0
24	55	55	54	54	54	53	53	52	52	52	51	51	50	50	50	49	49	48	48	1	0
25	58	57	57	56	56	55	55	55	54	54	53	53	52	52	52	51	51	50	50	1	0
26	60	59	59	58	58	58	57	57	56	56	55	55	55	54	54	53	53	52	52	1	0
27	62	62	61	61	60	60	59	59	58	58	58	57	57	56	56	55	55	54	54	1	0
28	64	64	63	63	63	62	62	61	61	60	60	59	59	58	58	57	57	56	56	1	0
29	67	66	66	65	65	64	64	63	63	62	62	61	61	60	60	59	59	58	58	1	0
30	69	68	68	68	67	66	66	66	65	64	64	64	63	62	62	62	61	60	60	**1**	0
31	71	71	70	70	69	69	68	68	67	67	66	66	65	65	64	64	63	63	62	1	1
32	74	73	73	72	71	71	70	70	69	69	68	68	67	67	66	66	65	65	64	1	1
33	76	75	75	74	74	73	73	72	72	71	70	70	69	69	68	68	67	67	66	1	1
34	78	78	77	76	76	75	75	74	74	73	73	72	71	71	70	70	69	69	68	1	1
35	80	80	79	79	78	78	77	76	76	75	75	74	74	73	72	72	71	71	70	1	1
36	83	82	82	81	80	80	79	79	78	77	77	76	76	75	74	74	73	73	72	1	1
37	85	84	84	83	83	82	81	81	80	80	79	78	78	77	76	76	75	75	74	1	1
38	87	87	86	86	85	84	84	83	82	82	81	80	80	79	79	78	77	77	76	1	1
39	90	89	88	88	87	86	86	85	84	84	83	83	82	81	81	80	79	79	78	1	1
40	92	91	91	90	89	89	88	87	87	86	85	85	84	83	83	82	81	81	80	1	1
41	94	94	93	92	92	91	90	90	89	88	87	87	86	85	85	84	83	83	82	1	1
42	97	96	95	94	94	93	92	92	91	90	90	89	88	88	87	86	85	85	84	1	1
43	99	98	97	97	96	95	95	94	93	92	92	91	90	90	89	88	87	87	86	1	1
44	101	100	100	99	98	98	97	96	95	95	94	93	92	92	91	90	89	89	88	1	1
45	104	103	102	101	100	100	99	98	98	97	96	95	94	94	93	92	92	91	90	2	1
46	106	105	104	104	103	102	101	100	100	99	98	97	97	96	95	94	94	93	92	2	1
47	108	107	107	106	105	104	103	103	102	101	100	99	99	98	97	96	96	95	94	2	1
48	110	110	109	108	107	106	106	105	104	103	102	102	101	100	99	98	98	97	96	2	1
49	113	112	111	110	109	109	108	107	106	105	105	104	103	102	101	100	100	99	98	2	1
50	115	114	113	112	112	111	110	109	108	108	107	106	105	104	103	102	102	101	100	2	1
51	117	116	116	115	114	113	112	111	110	110	109	108	107	106	105	105	104	103	102	2	1
52	120	119	118	117	116	115	114	114	113	112	111	110	109	108	107	107	106	105	104	2	1
53	122	121	120	119	118	117	117	116	115	114	113	112	111	110	110	109	108	107	106	2	1
54	124	123	122	122	121	120	119	118	117	116	115	114	113	112	112	111	110	109	108	2	1
55	126	126	125	124	123	122	121	120	119	118	117	116	116	115	114	113	112	111	110	2	1
56	129	128	127	126	125	124	123	122	121	120	119	119	118	117	116	115	114	113	112	2	1
57	131	130	129	128	127	126	125	124	124	123	122	121	120	119	118	117	116	115	114	2	1
58	133	132	131	130	130	129	128	127	126	125	124	123	122	121	120	119	118	117	116	2	1
59	136	135	134	133	132	131	130	129	128	127	126	125	124	123	122	121	120	119	118	2	1
60	138	137	136	135	134	133	132	131	130	129	128	127	126	125	124	123	122	121	120	**2**	**1**
″	**138**	**137**	**136**	**135**	**134**	**133**	**132**	**131**	**130**	**129**	**128**	**127**	**126**	**125**	**124**	**123**	**122**	**121**	**120**	**2**	**1**
	Proportional Parts																			

′	l sin 9.	d 1′	l csc 10.	l tan 9.	d 1′	l cot 10.	l sec 10.	d 1′	l cos 9.	′
0	01923	120	98077	02162	121	97838	00239	1	99761	60
1	02043	120	97957	283	121	717	240	1	760	59
2	163	120	837	404	121	596	241	2	759	58
3	283	119	717	525	120	475	243	1	757	57
4	402	118	598	645	121	355	244	1	756	56
5	520	119	480	766	119	234	245	2	755	55
6	639	118	361	885	120	115	247	1	753	54
7	757	117	243	03005	119	96995	248	1	752	53
8	874	118	126	124	118	876	249	2	751	52
9	992	117	008	242	119	758	251	1	749	51
10	03109	117	96891	361	118	639	252	1	748	50
11	226	116	774	479	118	521	253	2	747	49
12	342	116	658	597	117	403	255	1	745	48
13	458	116	542	714	118	286	256	2	744	47
14	574	116	426	832	116	168	258	1	742	46
15	690	115	310	948	117	052	259	1	741	45
16	805	115	195	04065	116	95935	260	2	740	44
17	920	114	080	181	116	819	262	1	738	43
18	04034	115	95966	297	116	703	263	1	737	42
19	149	113	851	413	115	587	264	2	736	41
20	262	114	738	528	115	472	266	1	734	40
21	376	114	624	643	115	357	267	2	733	39
22	490	113	510	758	115	242	269	1	731	38
23	603	112	397	873	114	127	270	2	730	37
24	715	113	285	987	114	013	272	1	728	36
25	828	112	172	05101	113	94899	273	1	727	35
26	940	112	060	214	114	786	274	2	726	34
27	05052	112	94948	328	113	672	276	1	724	33
28	164	111	836	441	112	559	277	2	723	32
29	275	111	725	553	113	447	279	1	721	31
30	05386	111	94614	05666	112	94334	00280	2	99720	30
31	497	110	503	778	112	222	282	1	718	29
32	607	110	393	890	112	110	283	1	717	28
33	717	110	283	06002	111	93998	284	2	716	27
34	827	110	173	113	111	887	286	1	714	26
35	937	109	063	224	111	776	287	2	713	25
36	06046	109	93954	335	110	665	289	1	711	24
37	155	109	845	445	111	555	290	2	710	23
38	264	108	736	556	110	444	292	1	708	22
39	372	109	628	666	109	334	293	2	707	21
40	481	108	519	775	110	225	295	1	705	20
41	589	107	411	885	109	115	296	2	704	19
42	696	108	304	994	109	006	298	1	702	18
43	804	107	196	07103	108	92897	299	2	701	17
44	911	107	089	211	109	789	301	1	699	16
45	07018	106	92982	320	108	680	302	2	698	15
46	124	107	876	428	108	572	304	1	696	14
47	231	106	769	536	107	464	305	2	695	13
48	337	105	663	643	108	357	307	1	693	12
49	442	106	558	751	107	249	308	2	692	11
50	548	105	452	858	106	142	310	1	690	10
51	653	105	347	964	107	036	311	2	689	9
52	758	105	242	08071	106	91929	313	1	687	8
53	863	105	137	177	106	823	314	2	686	7
54	968	104	032	283	106	717	316	1	684	6
55	08072	104	91928	389	106	611	317	2	683	5
56	176	104	824	495	105	505	319	1	681	4
57	280	103	720	600	105	400	320	2	680	3
58	383	103	617	705	105	295	322	1	678	2
59	486	103	514	810	104	190	323	2	677	1
60	08589		91411	08914		91086	00325		99675	0
′	9. l cos	d 1′	10. l sec	9. l cot	d 1′	10. l tan	10. l csc	d 1′	9. l sin	′

″	Proportional Parts 121	120	119	118	117
0	0	0	0	0	0
1	2	2	2	2	2
2	4	4	4	4	4
3	6	6	6	6	6
4	8	8	8	8	8
5	10	10	10	10	10
6	12	12	12	12	12
7	14	14	14	14	14
8	16	16	16	16	16
9	18	18	18	18	18
10	20	20	20	20	20
11	22	22	22	22	21
12	24	24	24	24	23
13	26	26	26	26	25
14	28	28	28	28	27
15	30	30	30	29	29
16	32	32	32	31	31
17	34	34	34	33	33
18	36	36	36	35	35
19	38	38	38	37	37
20	40	40	40	39	39
21	42	42	42	41	41
22	44	44	44	43	43
23	46	46	46	45	45
24	48	48	48	47	47
25	50	50	50	49	49
26	52	52	52	51	51
27	54	54	54	53	53
28	56	56	56	55	55
29	58	58	58	57	57
30	60	60	60	59	58
31	63	62	61	61	60
32	65	64	63	63	62
33	67	66	65	65	64
34	69	68	67	67	66
35	71	70	69	69	68
36	73	72	71	71	70
37	75	74	73	73	72
38	77	76	75	75	74
39	79	78	77	77	76
40	81	80	79	79	78
41	83	82	81	81	80
42	85	84	83	83	82
43	87	86	85	85	84
44	89	88	87	87	86
45	91	90	89	89	88
46	93	92	91	90	90
47	95	94	93	92	92
48	97	96	95	94	94
49	99	98	97	96	96
50	101	100	99	98	98
51	103	102	101	100	99
52	105	104	103	102	101
53	107	106	105	104	103
54	109	108	107	106	105
55	111	110	109	108	107
56	113	112	111	110	109
57	115	114	113	112	111
58	117	116	115	114	113
59	119	118	117	116	115
60	121	120	119	118	117
″	121 Proportional Parts	120	119	118	117

TABLE II

″	Proportional Parts													
	116	**115**	**114**	**113**	**112**	**111**	**110**	**109**	**108**	**107**	**106**	**105**	**104**	**2**	**1**
0	0	0	0	0	0	0	0	0	0	0	0	0	0	0	0
1	2	2	2	2	2	2	2	2	2	2	2	2	2	0	0
2	4	4	4	4	4	4	4	4	4	4	4	4	3	0	0
3	6	6	6	6	6	6	5	5	5	5	5	5	5	0	0
4	8	8	8	8	7	7	7	7	7	7	7	7	7	0	0
5	10	10	9	9	9	9	9	9	9	9	9	9	9	0	0
6	12	12	11	11	11	11	11	11	11	11	11	10	10	0	0
7	14	13	13	13	13	13	13	13	13	12	12	12	12	0	0
8	15	15	15	15	15	15	15	15	14	14	14	14	14	0	0
9	17	17	17	17	17	17	17	16	16	16	16	16	16	0	0
10	19	19	19	19	19	18	18	18	18	18	18	18	17	0	0
11	21	21	21	21	21	20	20	20	20	20	19	19	19	0	0
12	23	23	23	23	22	22	22	22	22	21	21	21	21	0	0
13	25	25	25	24	24	24	24	24	23	23	23	23	23	0	0
14	27	27	27	26	26	26	26	25	25	25	25	24	24	0	0
15	29	29	29	28	28	28	27	27	27	27	27	26	26	0	0
16	31	31	30	30	30	30	29	29	29	29	28	28	28	1	0
17	33	33	32	32	32	31	31	31	31	30	30	30	29	1	0
18	35	34	34	34	34	33	33	33	32	32	32	32	31	1	0
19	37	36	36	36	35	35	35	35	34	34	34	33	33	1	0
20	39	38	38	38	37	37	37	36	36	36	35	35	35	1	0
21	41	40	40	40	39	39	39	38	38	37	37	37	36	1	0
22	43	42	42	41	41	41	40	40	40	39	39	38	38	1	0
23	44	44	44	43	43	43	42	42	41	41	41	40	40	1	0
24	46	46	46	45	45	44	44	44	43	43	42	42	42	1	0
25	48	48	47	47	47	46	46	45	45	45	44	44	43	1	0
26	50	50	49	49	49	48	48	47	47	46	46	46	45	1	0
27	52	52	51	51	50	50	49	49	49	48	48	47	47	1	0
28	54	54	53	53	52	52	51	51	50	50	49	49	49	1	0
29	56	56	55	55	54	54	53	53	52	52	51	51	50	1	0
30	58	58	57	56	56	56	55	54	54	54	53	52	52	1	0
31	60	59	59	58	58	57	57	56	56	55	55	54	54	1	1
32	62	61	61	60	60	59	59	58	58	57	57	56	55	1	1
33	64	63	63	62	62	61	61	60	59	59	58	58	57	1	1
34	66	65	65	64	63	63	62	62	61	61	60	60	59	1	1
35	68	67	67	66	65	65	64	64	63	62	62	61	61	1	1
36	70	69	68	68	67	67	66	65	65	64	64	63	62	1	1
37	72	71	70	70	69	68	68	67	67	66	65	65	64	1	1
38	73	73	72	72	71	70	70	69	68	68	67	66	66	1	1
39	75	75	74	73	73	72	72	71	70	70	69	68	68	1	1
40	77	77	76	75	75	74	73	73	72	71	71	70	69	1	1
41	79	79	78	77	77	76	75	74	74	73	72	72	71	1	1
42	81	80	80	79	78	78	77	76	76	75	74	74	73	1	1
43	83	82	82	81	80	80	79	78	77	77	76	75	75	1	1
44	85	84	84	83	82	81	81	80	79	78	78	77	76	1	1
45	87	86	85	85	84	83	83	82	81	80	79	79	78	2	1
46	89	88	87	87	86	85	84	84	83	82	81	80	80	2	1
47	91	90	89	89	88	87	86	85	85	84	83	82	81	2	1
48	93	92	91	90	90	89	88	87	86	86	85	84	83	2	1
49	95	94	93	92	91	91	90	89	88	87	87	86	85	2	1
50	97	96	95	94	93	92	92	91	90	89	88	88	87	2	1
51	99	98	97	96	95	94	93	93	92	91	90	89	88	2	1
52	101	100	99	98	97	96	95	94	94	93	92	91	90	2	1
53	102	102	101	100	99	98	97	96	95	95	94	93	92	2	1
54	104	104	103	102	101	100	99	98	97	96	95	94	94	2	1
55	106	105	105	104	103	102	101	100	99	98	97	96	95	2	1
56	108	107	106	105	105	104	103	102	101	100	99	98	97	2	1
57	110	109	108	107	106	105	105	104	103	102	101	100	99	2	1
58	112	111	110	109	108	107	106	105	104	103	102	102	101	2	1
59	114	113	112	111	110	109	108	107	106	105	104	103	102	2	1
60	116	115	114	113	112	111	110	109	108	107	106	105	104	**2**	**1**
″	**116**	**115**	**114**	**113**	**112**	**111**	**110**	**109**	**108**	**107**	**106**	**105**	**104**	**2**	**1**
	Proportional Parts														

′	l sin 9.	d 1′	l csc 10.	l tan 9.	d 1′	l cot 10.	l sec 10.	d 1′	l cos 9.	′
0	08589	103	91411	08914	105	91086	00325	1	99675	60
1	692	103	308	09019	104	90981	326	2	674	59
2	795	102	205	123	104	877	328	2	672	58
3	897	102	103	227	103	773	330	1	670	57
4	999	102	001	330	104	670	331	2	669	56
5	09101	101	90899	434	103	566	333	1	667	55
6	202	102	798	537	103	463	334	2	666	54
7	304	101	696	640	102	360	336	1	664	53
8	405	101	595	742	103	258	337	2	663	52
9	506	100	494	845	102	155	339	2	661	51
10	606	101	394	947	102	053	341	1	659	50
11	707	100	293	10049	101	89951	342	2	658	49
12	807	100	193	150	102	850	344	1	656	48
13	907	99	093	252	101	748	345	2	655	47
14	10006	100	89994	353	101	647	347	2	653	46
15	106	99	894	454	101	546	349	1	651	45
16	205	99	795	555	101	445	350	2	650	44
17	304	98	696	656	100	344	352	1	648	43
18	402	99	598	756	100	244	353	2	647	42
19	501	98	499	856	100	144	355	2	645	41
20	599	98	401	956	100	044	357	1	643	40
21	697	98	303	11056	99	88944	358	2	642	39
22	795	98	205	155	99	845	360	2	640	38
23	893	97	107	254	99	746	362	1	638	37
24	990	97	010	353	99	647	363	2	637	36
25	11087	97	88913	452	99	548	365	2	635	35
26	184	97	816	551	98	449	367	1	633	34
27	281	96	719	649	98	351	368	2	632	33
28	377	97	623	747	98	253	370	1	630	32
29	474	96	526	845	98	155	371	2	629	31
30	11570	96	88430	11943	97	88057	00373	2	99627	30
31	666	95	334	12040	98	87960	375	1	625	29
32	761	96	239	138	97	862	376	2	624	28
33	857	95	143	235	97	765	378	2	622	27
34	952	95	048	332	96	668	380	2	620	26
35	12047	95	87953	428	97	572	382	1	618	25
36	142	94	858	525	96	475	383	2	617	24
37	236	95	764	621	96	379	385	2	615	23
38	331	94	669	717	96	283	387	1	613	22
39	425	94	575	813	96	187	388	2	612	21
40	519	93	481	909	95	091	390	2	610	20
41	612	94	388	13004	95	86996	392	1	608	19
42	706	93	294	099	95	901	393	2	607	18
43	799	93	201	194	95	806	395	2	605	17
44	892	93	108	289	95	711	397	2	603	16
45	985	93	015	384	94	616	399	1	601	15
46	13078	93	86922	478	95	522	400	2	600	14
47	171	92	829	573	94	427	402	2	598	13
48	263	92	737	667	94	333	404	1	596	12
49	355	92	645	761	93	239	405	2	595	11
50	447	92	553	854	94	146	407	2	593	10
51	539	91	461	948	93	052	409	2	591	9
52	630	92	370	14041	93	85959	411	1	589	8
53	722	91	278	134	93	866	412	2	588	7
54	813	91	187	227	93	773	414	2	586	6
55	904	90	096	320	92	680	416	2	584	5
56	994	91	006	412	92	588	418	1	582	4
57	14085	90	85915	504	93	496	419	2	581	3
58	175	91	825	597	91	403	421	2	579	2
59	266	90	734	688	92	312	423	2	577	1
60	14356		85644	14780		85220	00425		99575	0
′	9. l cos	d 1′	10. l sec	9. l cot	d 1′	10. l tan	10. l csc	d 1′	9. l sin	′

″	Proportional Parts 105	104	103	102
0	0	0	0	0
1	2	2	2	2
2	4	3	3	3
3	5	5	5	5
4	7	7	7	7
5	9	9	9	9
6	10	10	10	10
7	12	12	12	12
8	14	14	14	14
9	16	16	15	15
10	18	17	17	17
11	19	19	19	19
12	21	21	21	20
13	23	23	22	22
14	24	24	24	24
15	26	26	26	25
16	28	28	27	27
17	30	29	29	29
18	32	31	31	31
19	33	33	33	32
20	35	35	34	34
21	37	36	36	36
22	38	38	38	37
23	40	40	39	39
24	42	42	41	41
25	44	43	43	43
26	46	45	45	44
27	47	47	46	46
28	49	49	48	48
29	51	50	50	49
30	52	52	52	51
31	54	54	53	53
32	56	55	55	54
33	58	57	57	56
34	60	59	58	58
35	61	61	60	59
36	63	62	62	61
37	65	64	64	63
38	66	66	65	65
39	68	68	67	66
40	70	69	69	68
41	72	71	70	70
42	74	73	72	71
43	75	75	74	73
44	77	76	76	75
45	79	78	77	77
46	80	80	79	78
47	82	81	81	80
48	84	83	82	82
49	86	85	84	83
50	88	87	86	85
51	89	88	88	87
52	91	90	89	88
53	93	92	91	90
54	94	94	93	92
55	96	95	94	93
56	98	97	96	95
57	100	99	98	97
58	102	101	100	99
59	103	102	101	100
60	105	104	103	102
″	105 Proportional Parts	104	103	102

97° 82°

TABLE II

″	Proportional Parts												
	101	100	99	98	97	96	95	94	93	92	91	90	2	1
0	0	0	0	0	0	0	0	0	0	0	0	0	0	0
1	2	2	2	2	2	2	2	2	2	2	2	1	0	0
2	3	3	3	3	3	3	3	3	3	3	3	3	0	0
3	5	5	5	5	5	5	5	5	5	5	5	5	0	0
4	7	7	7	7	6	6	6	6	6	6	6	6	0	0
5	8	8	8	8	8	8	8	8	8	8	8	7	0	0
6	10	10	10	10	10	10	10	9	9	9	9	9	0	0
7	12	12	12	11	11	11	11	11	11	11	11	11	0	0
8	13	13	13	13	13	13	13	13	12	12	12	12	0	0
9	15	15	15	15	15	14	14	14	14	14	14	13	0	0
10	17	17	16	16	16	16	16	16	16	15	15	15	0	0
11	19	18	18	18	18	18	17	17	17	17	17	17	0	0
12	20	20	20	20	19	19	19	19	19	18	18	18	0	0
13	22	22	21	21	21	21	21	20	20	20	20	19	0	0
14	24	23	23	23	23	22	22	22	22	21	21	21	0	0
15	25	25	25	24	24	24	24	23	23	23	23	23	0	0
16	27	27	26	26	26	26	25	25	25	25	24	24	1	0
17	29	28	28	28	27	27	27	27	26	26	26	25	1	0
18	30	30	30	29	29	29	28	28	28	28	27	27	1	0
19	32	32	31	31	31	30	30	30	29	29	29	29	1	0
20	34	33	33	33	32	32	32	31	31	31	30	30	1	0
21	35	35	35	34	34	34	33	33	33	32	32	31	1	0
22	37	37	36	36	36	35	35	34	34	34	33	33	1	0
23	39	38	38	38	37	37	36	36	36	35	35	35	1	0
24	40	40	40	39	39	38	38	38	37	37	36	36	1	0
25	42	42	41	41	40	40	40	39	39	38	38	37	1	0
26	44	43	43	42	42	42	41	41	40	40	39	39	1	0
27	45	45	45	44	44	43	43	42	42	41	41	41	1	0
28	47	47	46	46	45	45	44	44	43	43	42	42	1	0
29	49	48	48	47	47	46	46	45	45	44	44	43	1	0
30	50	50	50	49	48	48	48	47	46	46	46	45	1	0
31	52	52	51	51	50	50	49	49	48	48	47	47	1	1
32	54	53	53	52	52	51	51	50	50	49	49	48	1	1
33	56	55	54	54	53	53	52	52	51	51	50	49	1	1
34	57	57	56	56	55	54	54	53	53	52	52	51	1	1
35	59	58	58	57	57	56	55	55	54	54	53	53	1	1
36	61	60	59	59	58	58	57	56	56	55	55	54	1	1
37	62	62	61	60	60	59	59	58	57	57	56	55	1	1
38	64	63	63	62	61	61	60	60	59	58	58	57	1	1
39	66	65	64	64	63	62	62	61	60	60	59	59	1	1
40	67	67	66	65	65	64	63	63	62	61	61	60	1	1
41	69	68	68	67	66	66	65	64	64	63	62	61	1	1
42	71	70	69	69	68	67	66	66	65	64	64	63	1	1
43	72	72	71	70	70	69	68	67	67	66	65	65	1	1
44	74	73	73	72	71	70	70	69	68	67	67	66	1	1
45	76	75	74	73	73	72	71	71	70	69	68	67	2	1
46	77	77	76	75	74	74	73	72	71	71	70	69	2	1
47	79	78	78	77	76	75	74	74	73	72	71	71	2	1
48	81	80	79	78	78	77	76	75	74	74	73	72	2	1
49	82	82	81	80	79	78	78	77	76	75	74	73	2	1
50	84	83	82	82	81	80	79	78	78	77	76	75	2	1
51	86	85	84	83	82	82	81	80	79	78	77	77	2	1
52	88	87	86	85	84	83	82	81	81	80	79	78	2	1
53	89	88	87	87	86	85	84	83	82	81	80	79	2	1
54	91	90	89	88	87	86	86	85	84	83	82	81	2	1
55	93	92	91	90	89	88	87	86	85	84	83	83	2	1
56	94	93	92	91	91	90	89	88	87	86	85	84	2	1
57	96	95	94	93	92	91	90	89	88	87	86	85	2	1
58	98	97	96	95	94	93	92	91	90	89	88	87	2	1
59	99	98	97	96	95	94	93	92	91	90	89	89	2	1
60	101	100	99	98	97	96	95	94	93	92	91	90	2	1
″	101	100	99	98	97	96	95	94	93	92	91	90	2	1
	Proportional Parts													

′	*l* sin 9.	d 1′	*l* csc 10.	*l* tan 9.	d 1′	*l* cot 10.	*l* sec 10.	d 1′	*l* cos 9.	′
0	14356	89	85644	14780	92	85220	00425	1	99575	60
1	445	90	555	872	91	128	426	2	574	59
2	535	89	465	963	91	037	428	2	572	58
3	624	90	376	15054	91	84946	430	2	570	57
4	714	89	286	145	91	855	432	2	568	56
5	803	88	197	236	91	764	434	1	566	55
6	891	89	109	327	90	673	435	2	565	54
7	980	89	020	417	91	583	437	2	563	53
8	15069	88	84931	508	90	492	439	2	561	52
9	157	88	843	598	90	402	441	2	559	51
10	245	88	755	688	89	312	443	1	557	50
11	333	88	667	777	90	223	444	2	556	49
12	421	87	579	867	89	133	446	2	554	48
13	508	88	492	956	90	044	448	2	552	47
14	596	87	404	16046	89	83954	450	2	550	46
15	683	87	317	135	89	865	452	2	548	45
16	770	87	230	224	88	776	454	1	546	44
17	857	87	143	312	89	688	455	2	545	43
18	944	86	056	401	88	599	457	2	543	42
19	16030	86	83970	489	88	511	459	2	541	41
20	116	87	884	577	88	423	461	2	539	40
21	203	86	797	665	88	335	463	2	537	39
22	289	85	711	753	88	247	465	2	535	38
23	374	86	626	841	87	159	467	1	533	37
24	460	85	540	928	88	072	468	2	532	36
25	545	86	455	17016	87	82984	470	2	530	35
26	631	85	369	103	87	897	472	2	528	34
27	716	85	284	190	87	810	474	2	526	33
28	801	85	199	277	86	723	476	2	524	32
29	886	84	114	363	87	637	478	2	522	31
30	16970	85	83030	17450	86	82550	00480	2	99520	30
31	17055	84	82945	536	86	464	482	1	518	29
32	139	84	861	622	86	378	483	2	517	28
33	223	84	777	708	86	292	485	2	515	27
34	307	84	693	794	86	206	487	2	513	26
35	391	83	609	880	85	120	489	2	511	25
36	474	84	526	965	86	035	491	2	509	24
37	558	83	442	18051	85	81949	493	2	507	23
38	641	83	359	136	85	864	495	2	505	22
39	724	83	276	221	85	779	497	2	503	21
40	807	83	193	306	85	694	499	2	501	20
41	890	83	110	391	84	609	501	2	499	19
42	973	82	027	475	85	525	503	2	497	18
43	18055	82	81945	560	84	440	505	1	495	17
44	137	83	863	644	84	356	506	2	494	16
45	220	82	780	728	84	272	508	2	492	15
46	302	81	698	812	84	188	510	2	490	14
47	383	82	617	896	83	104	512	2	488	13
48	465	82	535	979	84	021	514	2	486	12
49	547	81	453	19063	83	80937	516	2	484	11
50	628	81	372	146	83	854	518	2	482	10
51	709	81	291	229	83	771	520	2	480	9
52	790	81	210	312	83	688	522	2	478	8
53	871	81	129	395	83	605	524	2	476	7
54	952	81	048	478	83	522	526	2	474	6
55	19033	80	80967	561	82	439	528	2	472	5
56	113	80	887	643	82	357	530	2	470	4
57	193	80	807	725	82	275	532	2	468	3
58	273	80	727	807	82	193	534	2	466	2
59	353	80	647	889	82	111	536	2	464	1
60	19433		80567	19971		80029	00538		99462	0
′	9. *l* cos	d 1′	10. *l* sec	9. *l* cot	d 1′	10. *l* tan	10. *l* csc	d 1′	9. *l* sin	′

″	Proportional Parts 92	91	90
0	0	0	0
1	2	2	1
2	3	3	3
3	5	5	5
4	6	6	6
5	8	8	7
6	9	9	9
7	11	11	11
8	12	12	12
9	14	14	13
10	15	15	15
11	17	17	17
12	18	18	18
13	20	20	19
14	21	21	21
15	23	23	23
16	25	24	24
17	26	26	25
18	28	27	27
19	29	29	29
20	31	30	30
21	32	32	31
22	34	33	33
23	35	35	35
24	37	36	36
25	38	38	37
26	40	39	39
27	41	41	41
28	43	42	42
29	44	44	43
30	46	46	45
31	48	47	47
32	49	49	48
33	51	50	49
34	52	52	51
35	54	53	53
36	55	55	54
37	57	56	55
38	58	58	57
39	60	59	59
40	61	61	60
41	63	62	61
42	64	64	63
43	66	65	65
44	67	67	66
45	69	68	67
46	71	70	69
47	72	71	71
48	74	73	72
49	75	74	73
50	77	76	75
51	78	77	77
52	80	79	78
53	81	80	79
54	83	82	81
55	84	83	83
56	86	85	84
57	87	86	85
58	89	88	87
59	90	89	89
60	92	91	90
″	92 Proportional Parts	91	90

TABLE II

″	Proportional Parts											
	89	88	87	86	85	84	83	82	81	80	2	1
0	0	0	0	0	0	0	0	0	0	0	0	0
1	1	1	1	1	1	1	1	1	1	1	0	0
2	3	3	3	3	3	3	3	3	3	3	0	0
3	4	4	4	4	4	4	4	4	4	4	0	0
4	6	6	6	6	6	6	6	5	5	5	0	0
5	7	7	7	7	7	7	7	7	7	7	0	0
6	9	9	9	9	8	8	8	8	8	8	0	0
7	10	10	10	10	10	10	10	10	9	9	0	0
8	12	12	12	11	11	11	11	11	11	11	0	0
9	13	13	13	13	13	13	12	12	12	12	0	0
10	15	15	14	14	14	14	14	14	14	13	0	0
11	16	16	16	16	16	15	15	15	15	15	0	0
12	18	18	17	17	17	17	17	16	16	16	0	0
13	19	19	19	19	18	18	18	18	18	17	0	0
14	21	21	20	20	20	20	19	19	19	19	0	0
15	22	22	22	21	21	21	21	21	20	20	0	0
16	24	23	23	23	23	22	22	22	22	21	1	0
17	25	25	25	24	24	24	24	23	23	23	1	0
18	27	26	26	26	26	25	25	25	24	24	1	0
19	28	28	28	27	27	27	26	26	26	25	1	0
20	30	29	29	29	28	28	28	27	27	27	1	0
21	31	31	30	30	30	29	29	29	28	28	1	0
22	33	32	32	32	31	31	30	30	30	29	1	0
23	34	34	33	33	33	32	32	31	31	31	1	0
24	36	35	35	34	34	34	33	33	32	32	1	0
25	37	37	36	36	35	35	35	34	34	33	1	0
26	39	38	38	37	37	36	36	36	35	35	1	0
27	40	40	39	39	38	38	37	37	36	36	1	0
28	42	41	41	40	40	39	39	38	38	37	1	0
29	43	43	42	42	41	41	40	40	39	39	1	0
30	44	44	44	43	42	42	42	41	40	40	1	0
31	46	45	45	44	44	43	43	42	42	41	1	1
32	47	47	46	46	45	45	44	44	43	43	1	1
33	49	48	48	47	47	46	46	45	45	44	1	1
34	50	50	49	49	48	48	47	46	46	45	1	1
35	52	51	51	50	50	49	48	48	47	47	1	1
36	53	53	52	52	51	50	50	49	49	48	1	1
37	55	54	54	53	52	52	51	51	50	49	1	1
38	56	56	55	54	54	53	53	52	51	51	1	1
39	58	57	57	56	55	55	54	53	53	52	1	1
40	59	59	58	57	57	56	55	55	54	53	1	1
41	61	60	59	59	58	57	57	56	55	55	1	1
42	62	62	61	60	60	59	58	57	57	56	1	1
43	64	63	62	62	61	60	59	59	58	57	1	1
44	65	65	64	63	62	62	61	60	59	59	1	1
45	67	66	65	65	64	63	62	61	61	60	2	1
46	68	67	67	66	65	64	64	63	62	61	2	1
47	70	69	68	67	67	66	65	64	63	63	2	1
48	71	70	70	69	68	67	66	66	65	64	2	1
49	73	72	71	70	69	69	68	67	66	65	2	1
50	74	73	72	72	71	70	69	68	68	67	2	1
51	76	75	74	73	72	71	71	70	69	68	2	1
52	77	76	75	75	74	73	72	71	70	69	2	1
53	79	78	77	76	75	74	73	72	72	71	2	1
54	80	79	78	77	76	76	75	74	73	72	2	1
55	82	81	80	79	78	77	76	75	74	73	2	1
56	83	82	81	80	79	78	77	77	76	75	2	1
57	85	84	83	82	81	80	79	78	77	76	2	1
58	86	85	84	83	82	81	80	79	78	77	2	1
59	88	87	86	85	84	83	82	81	80	79	2	1
60	89	88	87	86	85	84	83	82	81	80	2	1
″	89	88	87	86	85	84	83	82	81	80	2	1
	Proportional Parts											

9° TABLE II 170°

′	l sin 9.	d 1′	l csc 10.	l tan 9.	d 1′	l cot 10.	l sec 10.	d 1′	l cos 9.	′
0	19433	80	80567	19971	82	80029	00538	2	99462	60
1	513	79	487	20053	81	79947	540	2	460	59
2	592	80	408	134	82	866	542	2	458	58
3	672	79	328	216	81	784	544	2	456	57
4	751	79	249	297	81	703	546	2	454	56
5	830	79	170	378	81	622	548	2	452	55
6	909	79	091	459	81	541	550	2	450	54
7	988	79	012	540	81	460	552	2	448	53
8	20067	78	79933	621	80	379	554	2	446	52
9	145	78	855	701	81	299	556	2	444	51
10	223	79	777	782	80	218	558	2	442	50
11	302	78	698	862	80	138	560	2	440	49
12	380	78	620	942	80	058	562	2	438	48
13	458	77	542	21022	80	78978	564	2	436	47
14	535	78	465	102	80	898	566	2	434	46
15	613	78	387	182	79	818	568	3	432	45
16	691	77	309	261	80	739	571	2	429	44
17	768	77	232	341	79	659	573	2	427	43
18	845	77	155	420	79	580	575	2	425	42
19	922	77	078	499	79	501	577	2	423	41
20	999	77	001	578	79	422	579	2	421	40
21	21076	77	78924	657	79	343	581	2	419	39
22	153	76	847	736	78	264	583	2	417	38
23	229	77	771	814	79	186	585	2	415	37
24	306	76	694	893	78	107	587	2	413	36
25	382	76	618	971	78	029	589	2	411	35
26	458	76	542	22049	78	77951	591	2	409	34
27	534	76	466	127	78	873	593	3	407	33
28	610	75	390	205	78	795	596	2	404	32
29	685	76	315	283	78	717	598	2	402	31
30	21761	75	78239	22361	77	77639	00600	2	99400	30
31	836	76	164	438	78	562	602	2	398	29
32	912	75	088	516	77	484	604	2	396	28
33	987	75	013	593	77	407	606	2	394	27
34	22062	75	77938	670	77	330	608	2	392	26
35	137	74	863	747	77	253	610	2	390	25
36	211	75	789	824	77	176	612	3	388	24
37	286	75	714	901	76	099	615	2	385	23
38	361	74	639	977	77	023	617	2	383	22
39	435	74	565	23054	76	76946	619	2	381	21
40	509	74	491	130	76	870	621	2	379	20
41	583	74	417	206	77	794	623	2	377	19
42	657	74	343	283	76	717	625	3	375	18
43	731	74	269	359	76	641	628	2	372	17
44	805	73	195	435	75	565	630	2	370	16
45	878	74	122	510	76	490	632	2	368	15
46	952	73	048	586	75	414	634	2	366	14
47	23025	73	76975	661	76	339	636	2	364	13
48	098	73	902	737	75	263	638	3	362	12
49	171	73	829	812	75	188	641	2	359	11
50	244	73	756	887	75	113	643	2	357	10
51	317	73	683	962	75	038	645	2	355	9
52	390	72	610	24037	75	75963	647	2	353	8
53	462	73	538	112	74	888	649	3	351	7
54	535	72	465	186	75	814	652	2	348	6
55	607	72	393	261	74	739	654	2	346	5
56	679	73	321	335	75	665	656	2	344	4
57	752	71	248	410	74	590	658	2	342	3
58	823	72	177	484	74	516	660	3	340	2
59	895	72	105	558	74	442	663	2	337	1
60	23967		76033	24632		75368	00665		99335	0
′	9. l cos	d 1′	10. l sec	9. l cot	d 1′	10. l tan	10. l csc	d 1′	9. l sin	′

99° 80°

″	82	81	80	79	78	77	76	75	74	73	72	71	3	2
	Proportional Parts													
0	0	0	0	0	0	0	0	0	0	0	0	0	0	0
1	1	1	1	1	1	1	1	1	1	1	1	1	0	0
2	3	3	3	3	3	3	3	2	2	2	2	2	0	0
3	4	4	4	4	4	4	4	4	4	4	4	4	0	0
4	5	5	5	5	5	5	5	5	5	5	5	5	0	0
5	7	7	7	7	7	6	6	6	6	6	6	6	0	0
6	8	8	8	8	8	8	8	8	7	7	7	7	0	0
7	10	9	9	9	9	9	9	9	9	9	8	8	0	0
8	11	11	11	11	10	10	10	10	10	10	10	9	0	0
9	12	12	12	12	12	12	11	11	11	11	11	11	0	0
10	14	14	13	13	13	13	13	12	12	12	12	12	0	0
11	15	15	15	14	14	14	14	14	14	13	13	13	1	0
12	16	16	16	16	16	15	15	15	15	15	14	14	1	0
13	18	18	17	17	17	17	16	16	16	16	16	15	1	0
14	19	19	19	18	18	18	18	18	17	17	17	17	1	0
15	21	20	20	20	19	19	19	19	19	18	18	18	1	0
16	22	22	21	21	21	21	20	20	20	19	19	19	1	1
17	23	23	23	22	22	22	22	21	21	21	20	20	1	1
18	25	24	24	24	23	23	23	22	22	22	22	21	1	1
19	26	26	25	25	25	24	24	24	23	23	23	22	1	1
20	27	27	27	26	26	26	25	25	25	24	24	24	1	1
21	29	28	28	28	27	27	27	26	26	26	25	25	1	1
22	30	30	29	29	29	28	28	28	27	27	26	26	1	1
23	31	31	31	30	30	30	29	29	28	28	28	27	1	1
24	33	32	32	32	31	31	30	30	30	29	29	28	1	1
25	34	34	33	33	33	32	32	31	31	30	30	30	1	1
26	36	35	35	34	34	33	33	32	32	32	31	31	1	1
27	37	36	36	36	35	35	34	34	33	33	32	32	1	1
28	38	38	37	37	36	36	35	35	35	34	34	33	1	1
29	40	39	39	38	38	37	37	36	36	35	35	34	1	1
30	41	40	40	40	39	38	38	38	37	36	36	36	2	1
31	42	42	41	41	40	40	39	39	38	38	37	37	2	1
32	44	43	43	42	42	41	41	40	39	39	38	38	2	1
33	45	45	44	43	43	42	42	41	41	40	40	39	2	1
34	46	46	45	45	44	44	43	42	42	41	41	40	2	1
35	48	47	47	46	45	45	44	44	43	43	42	41	2	1
36	49	49	48	47	47	46	46	45	44	44	43	43	2	1
37	51	50	49	49	48	47	47	46	46	45	44	44	2	1
38	52	51	51	50	49	49	48	48	47	46	46	45	2	1
39	53	53	52	51	51	50	49	49	48	47	47	46	2	1
40	55	54	53	53	52	51	51	50	49	49	48	47	2	1
41	56	55	55	54	53	53	52	51	51	50	49	49	2	1
42	57	57	56	55	55	54	53	52	52	51	50	50	2	1
43	59	58	57	57	56	55	54	54	53	52	52	51	2	1
44	60	59	59	58	57	56	56	55	54	54	53	52	2	1
45	61	61	60	59	59	58	57	56	55	55	54	53	2	2
46	63	62	61	61	60	59	58	58	57	56	55	54	2	2
47	64	63	63	62	61	60	60	59	58	57	56	56	2	2
48	66	65	64	63	62	62	61	60	59	58	58	57	2	2
49	67	66	65	65	64	63	62	61	60	60	59	58	2	2
50	68	68	67	66	65	64	63	62	62	61	60	59	2	2
51	70	69	68	67	66	65	65	64	63	62	61	60	3	2
52	71	70	69	68	68	67	66	65	64	63	62	62	3	2
53	72	72	71	70	69	68	67	66	65	64	64	63	3	2
54	74	73	72	71	70	69	68	68	67	66	65	64	3	2
55	75	74	73	72	71	71	70	69	68	67	66	65	3	2
56	77	76	75	74	73	72	71	70	69	68	67	66	3	2
57	78	77	76	75	74	73	72	71	70	69	68	67	3	2
58	79	78	77	76	75	74	73	72	72	71	70	69	3	2
59	81	80	79	78	77	76	75	74	73	72	71	70	3	2
60	82	81	80	79	78	77	76	75	74	73	72	71	3	2
″	82	81	80	79	78	77	76	75	74	73	72	71	3	2
	Proportional Parts													

′	*l* sin 9.	d 1′	*l* csc 10.	*l* tan 9.	d 1′	*l* cot 10.	*l* sec 10.	d 1′	*l* cos 9.	′
0	**23967**		**76033**	**24632**		**75368**	**00665**		**99335**	**60**
1	**24039**	72	**75961**	706	74	294	667	2	333	59
2	110	71	890	779	73	221	669	2	331	58
3	181	71	819	853	74	147	672	3	328	57
4	253	72	747	926	73	074	674	2	326	56
5	324	71	676	**25000**	74	000	676	2	324	**55**
6	395	71	605	073	73	**74927**	678	2	322	54
7	466	71	534	146	73	854	681	3	319	53
8	536	70	464	219	73	781	683	2	317	52
9	607	71	393	292	73	708	685	2	315	51
10	677	70	323	365	73	635	687	2	313	**50**
11	748	71	252	437	72	563	690	3	310	49
12	818	70	182	510	73	490	692	2	308	48
13	888	70	112	582	72	418	694	2	306	47
14	958	70	042	655	73	345	696	2	304	46
15	**25028**	70	**74972**	727	72	273	699	3	301	**45**
16	098	70	902	799	72	201	701	2	299	44
17	168	70	832	871	72	129	703	2	297	43
18	237	69	763	943	72	057	706	3	294	42
19	307	70	693	**26015**	72	**73985**	708	2	292	41
20	376	69	624	086	71	914	710	2	290	**40**
21	445	69	555	158	72	842	712	2	288	39
22	514	69	486	229	71	771	715	3	285	38
23	583	69	417	301	72	699	717	2	283	37
24	652	69	348	372	71	628	719	2	281	36
25	721	69	279	443	71	557	722	3	278	**35**
26	790	69	210	514	71	486	724	2	276	34
27	858	68	142	585	71	415	726	2	274	33
28	927	69	073	655	70	345	729	3	271	32
29	995	68	005	726	71	274	731	2	269	31
30	**26063**	68	**73937**	**26797**	71	**73203**	**00733**	2	**99267**	**30**
31	131	68	869	867	70	133	736	3	264	29
32	199	68	801	937	70	063	738	2	262	28
33	267	68	733	**27008**	71	**72992**	740	2	260	27
34	335	68	665	078	70	922	743	3	257	26
35	403	68	597	148	70	852	745	2	255	**25**
36	470	67	530	218	70	782	748	3	252	24
37	538	68	462	288	70	712	750	2	250	23
38	605	67	395	357	69	643	752	2	248	22
39	672	67	328	427	70	573	755	3	245	21
40	739	67	261	496	69	504	757	2	243	**20**
41	806	67	194	566	70	434	759	2	241	19
42	873	67	127	635	69	365	762	3	238	18
43	940	67	060	704	69	296	764	2	236	17
44	**27007**	67	**72993**	773	69	227	767	3	233	16
45	073	66	927	842	69	158	769	2	231	**15**
46	140	67	860	911	69	089	771	2	229	14
47	206	66	794	980	69	020	774	3	226	13
48	273	67	727	**28049**	69	**71951**	776	2	224	12
49	339	66	661	117	68	883	779	3	221	11
50	405	66	595	186	69	814	781	2	219	**10**
51	471	66	529	254	68	746	783	2	217	9
52	537	66	463	323	69	677	786	3	214	8
53	602	65	398	391	68	609	788	2	212	7
54	668	66	332	459	68	541	791	3	209	6
55	734	66	266	527	68	473	793	2	207	**5**
56	799	65	201	595	68	405	796	3	204	4
57	864	65	136	662	67	338	798	2	202	3
58	930	66	070	730	68	270	800	2	200	2
59	995	65	005	798	68	202	803	3	197	1
60	**28060**	65	**71940**	**28865**	67	**71135**	**00805**	2	**99195**	**0**
′	9. *l* cos	d 1′	10. *l* sec	9. *l* cot	d 1′	10. *l* tan	10. *l* csc	d 1′	9. *l* sin	′

″	Proportional Parts											
″	**74**	**73**	**72**	**71**	**70**	**69**	**68**	**67**	**66**	**65**	**3**	**2**
0	0	0	0	0	0	0	0	0	0	0	0	0
1	1	1	1	1	1	1	1	1	1	1	0	0
2	**2**	**2**	**2**	2	2	2	2	2	2	2	0	0
3	4	4	4	**4**	**3**	**3**	3	3	3	3	0	0
4	5	5	5	5	5	5	**5**	**4**	4	4	0	0
5	6	6	6	6	6	6	6	6	**5**	5	0	0
6	**7**	7	7	7	7	7	7	7	7	**6**	0	0
7	9	**9**	8	8	8	8	8	8	8	8	0	0
8	10	10	**10**	**9**	9	9	9	9	9	9	0	0
9	11	11	11	11	**11**	10	10	10	10	10	0	0
10	12	12	12	12	12	**12**	11	11	11	11	0	0
11	**14**	13	13	13	13	13	**12**	12	12	12	0	0
12	15	**15**	**14**	14	14	14	14	13	13	13	1	0
13	16	16	16	15	15	15	15	**15**	14	14	1	0
14	17	17	17	**17**	16	16	16	16	15	15	1	0
15	**19**	18	18	18	**17**	17	17	17	**17**	16	1	0
16	20	**19**	19	19	19	18	18	18	18	17	1	1
17	21	21	20	20	20	**20**	19	19	19	18	1	1
18	22	22	**22**	21	21	21	20	20	20	**20**	1	1
19	**23**	23	23	**22**	22	22	**22**	21	21	21	1	1
20	25	24	24	24	23	23	23	22	22	22	**1**	1
21	26	**26**	25	25	**25**	24	24	**23**	23	23	1	1
22	27	27	**26**	26	26	25	25	25	24	24	1	1
23	28	28	28	27	27	**26**	26	26	25	25	1	1
24	**30**	29	29	28	28	28	27	27	26	26	1	1
25	31	**30**	30	**30**	29	29	28	28	**27**	27	1	1
26	32	32	31	31	30	30	**29**	29	29	28	1	1
27	33	33	32	32	**31**	31	31	30	30	29	1	1
28	**35**	34	**34**	33	33	32	32	31	31	30	1	1
29	36	35	35	34	34	33	33	32	32	31	1	1
30	37	**36**	36	**36**	35	**34**	34	**34**	33	**32**	2	**1**
31	38	38	37	37	36	36	35	35	34	34	2	1
32	**39**	39	**38**	38	37	37	36	36	35	35	2	1
33	41	40	40	39	**39**	38	37	37	36	36	2	1
34	42	41	41	40	40	39	**39**	38	37	37	2	1
35	43	**43**	42	**41**	41	40	40	39	**39**	38	2	1
36	**44**	44	43	43	42	41	41	40	40	39	2	1
37	46	45	44	44	43	**43**	42	41	41	40	2	1
38	47	46	**46**	45	44	44	43	42	42	41	2	1
39	48	**47**	47	46	**45**	45	44	**44**	43	42	2	1
40	49	49	48	47	47	46	45	45	44	43	**2**	1
41	**51**	50	49	**49**	48	47	**46**	46	45	44	2	1
42	52	51	**50**	50	49	48	48	47	46	**46**	2	1
43	53	52	52	51	50	**49**	49	48	47	47	2	1
44	54	**54**	53	52	51	51	50	49	48	48	2	1
45	**55**	55	54	53	**53**	52	51	50	**49**	49	2	2
46	57	56	55	**54**	54	53	52	51	51	50	2	2
47	58	57	56	56	55	54	53	**52**	52	51	2	2
48	59	**58**	**58**	57	56	55	54	54	53	52	2	2
49	**60**	60	59	58	57	56	**56**	55	54	53	2	2
50	62	61	60	59	58	**58**	57	56	55	54	2	2
51	63	62	61	60	**59**	59	58	57	56	55	3	2
52	64	63	**62**	**62**	61	60	59	58	57	56	3	2
53	65	**64**	64	63	62	61	60	59	58	57	3	2
54	**67**	66	65	64	63	62	61	60	59	**58**	3	2
55	68	67	66	65	64	63	62	61	**61**	60	3	2
56	69	68	67	66	65	64	**63**	**63**	62	61	3	2
57	70	69	68	**67**	**67**	**66**	65	64	63	62	3	2
58	**72**	**71**	**70**	69	68	67	66	65	64	63	3	2
59	73	72	71	70	69	68	67	66	65	64	3	2
60	74	73	72	71	70	69	68	67	66	65	**3**	**2**
″	**74**	**73**	**72**	**71**	**70**	**69**	**68**	**67**	**66**	**65**	**3**	**2**
	Proportional Parts											

′	*l* sin 9.	d 1′	*l* csc 10.	*l* tan 9.	d 1′	*l* cot 10.	*l* sec 10.	d 1′	*l* cos 9.	′
0	28060		71940	28865		71135	00805		99195	60
1	125	65	875	933	68	067	808	3	192	59
2	190	65	810	29000	67	000	810	2	190	58
3	254	64	746	067	67	70933	813	3	187	57
4	319	65	681	134	67	866	815	2	185	56
5	384	65	616	201	67	799	818	3	182	55
6	448	64	552	268	67	732	820	2	180	54
7	512	64	488	335	67	665	823	3	177	53
8	577	65	423	402	67	598	825	2	175	52
9	641	64	359	468	66	532	828	3	172	51
10	705	64	295	535	67	465	830	2	170	50
11	769	64	231	601	66	399	833	3	167	49
12	833	64	167	668	67	332	835	2	165	48
13	896	63	104	734	66	266	838	3	162	47
14	960	64	040	800	66	200	840	2	160	46
15	29024	64	70976	866	66	134	843	3	157	45
16	087	63	913	932	66	068	845	2	155	44
17	150	63	850	998	66	002	848	3	152	43
18	214	64	786	30064	66	69936	850	2	150	42
19	277	63	723	130	66	870	853	3	147	41
20	340	63	660	195	65	805	855	2	145	40
21	403	63	597	261	66	739	858	3	142	39
22	466	63	534	326	65	674	860	2	140	38
23	529	63	471	391	65	609	863	3	137	37
24	591	62	409	457	66	543	865	2	135	36
25	654	63	346	522	65	478	868	3	132	35
26	716	62	284	587	65	413	870	2	130	34
27	779	63	221	652	65	348	873	3	127	33
28	841	62	159	717	65	283	876	3	124	32
29	903	62	097	782	65	218	878	2	122	31
30	29966	63	70034	30846	64	69154	00881	3	99119	30
31	30028	62	69972	911	65	089	883	2	117	29
32	090	62	910	975	64	025	886	3	114	28
33	151	61	849	31040	65	68960	888	2	112	27
34	213	62	787	104	64	896	891	3	109	26
35	275	62	725	168	64	832	894	3	106	25
36	336	61	664	233	65	767	896	2	104	24
37	398	62	602	297	64	703	899	3	101	23
38	459	61	541	361	64	639	901	2	099	22
39	521	62	479	425	64	575	904	3	096	21
40	582	61	418	489	64	511	907	3	093	20
41	643	61	357	552	63	448	909	2	091	19
42	704	61	296	616	64	384	912	3	088	18
43	765	61	235	679	63	321	914	2	086	17
44	826	61	174	743	64	257	917	3	083	16
45	887	61	113	806	63	194	920	3	080	15
46	947	60	053	870	64	130	922	2	078	14
47	31008	61	68992	933	63	067	925	3	075	13
48	068	60	932	996	63	004	928	3	072	12
49	129	61	871	32059	63	67941	930	2	070	11
50	189	60	811	122	63	878	933	3	067	10
51	250	61	750	185	63	815	936	3	064	9
52	310	60	690	248	63	752	938	2	062	8
53	370	60	630	311	63	689	941	3	059	7
54	430	60	570	373	62	627	944	3	056	6
55	490	60	510	436	63	564	946	2	054	5
56	549	59	451	498	62	502	949	3	051	4
57	609	60	391	561	63	439	952	3	048	3
58	669	60	331	623	62	377	954	2	046	2
59	728	59	272	685	62	315	957	3	043	1
60	31788	60	68212	32747	62	67253	00960	3	99040	0
′	9. *l* cos	d 1′	10. *l* sec	9. *l* cot	d 1′	10. *l* tan	10. *l* csc	d 1′	9. *l* sin	′

″	Proportional Parts 68	67	66	65	64	63	62	61	60	59	3	2
0	0	0	0	0	0	0	0	0	0	0	0	0
1	1	1	1	1	1	1	1	1	1	1	0	0
2	2	2	2	2	2	2	2	2	2	2	0	0
3	3	3	3	3	3	3	3	3	3	3	0	0
4	5	4	4	4	4	4	4	4	4	4	0	0
5	6	6	5	5	5	5	5	5	5	5	0	0
6	7	7	7	6	6	6	6	6	6	6	0	0
7	8	8	8	8	7	7	7	7	7	7	0	0
8	9	9	9	9	9	8	8	8	8	8	0	0
9	10	10	10	10	10	9	9	9	9	9	0	0
10	11	11	11	11	11	10	10	10	10	10	0	0
11	12	12	12	12	12	12	11	11	11	11	1	0
12	14	13	13	13	13	13	12	12	12	12	1	0
13	15	15	14	14	14	14	13	13	13	13	1	0
14	16	16	15	15	15	15	14	14	14	14	1	0
15	17	17	17	16	16	16	15	15	15	15	1	0
16	18	18	18	17	17	17	17	16	16	16	1	1
17	19	19	19	18	18	18	18	17	17	17	1	1
18	20	20	20	20	19	19	19	18	18	18	1	1
19	22	21	21	21	20	20	20	19	19	19	1	1
20	23	22	22	22	21	21	21	20	20	20	1	1
21	24	23	23	23	22	22	22	21	21	21	1	1
22	25	25	24	24	23	23	23	22	22	22	1	1
23	26	26	25	25	25	24	24	23	23	23	1	1
24	27	27	26	26	26	25	25	24	24	24	1	1
25	28	28	27	27	27	26	26	25	25	25	1	1
26	29	29	29	28	28	27	27	26	26	26	1	1
27	31	30	30	29	29	28	28	27	27	27	1	1
28	32	31	31	30	30	29	29	28	28	28	1	1
29	33	32	32	31	31	30	30	29	29	29	1	1
30	34	34	33	32	32	32	31	30	30	30	2	1
31	35	35	34	34	33	33	32	32	31	30	2	1
32	36	36	35	35	34	34	33	33	32	31	2	1
33	37	37	36	36	35	35	34	34	33	32	2	1
34	39	38	37	37	36	36	35	35	34	33	2	1
35	40	39	39	38	37	37	36	36	35	34	2	1
36	41	40	40	39	38	38	37	37	36	35	2	1
37	42	41	41	40	39	39	38	38	37	36	2	1
38	43	42	42	41	41	40	39	39	38	37	2	1
39	44	44	43	42	42	41	40	40	39	38	2	1
40	45	45	44	43	43	42	41	41	40	39	2	1
41	46	46	45	44	44	43	42	42	41	40	2	1
42	48	47	46	46	45	44	43	43	42	41	2	1
43	49	48	47	47	46	45	44	44	43	42	2	1
44	50	49	48	48	47	46	45	45	44	43	2	1
45	51	50	49	49	48	47	47	46	45	44	2	2
46	52	51	51	50	49	48	48	47	46	45	2	2
47	53	52	52	51	50	49	49	48	47	46	2	2
48	54	54	53	52	51	50	50	49	48	47	2	2
49	56	55	54	53	52	51	51	50	49	48	2	2
50	57	56	55	54	53	52	52	51	50	49	2	2
51	58	57	56	55	54	54	53	52	51	50	3	2
52	59	58	57	56	55	55	54	53	52	51	3	2
53	60	59	58	57	57	56	55	54	53	52	3	2
54	61	60	59	58	58	57	56	55	54	53	3	2
55	62	61	61	60	59	58	57	56	55	54	3	2
56	63	63	62	61	60	59	58	57	56	55	3	2
57	65	64	63	62	61	60	59	58	57	56	3	2
58	66	65	64	63	62	61	60	59	58	57	3	2
59	67	66	65	64	63	62	61	60	59	58	3	2
60	68	67	66	65	64	63	62	61	60	59	3	2
″	68	67	66	65	64	63	62	61	60	59	3	2

Proportional Parts

′	*l* sin 9.	d 1′	*l* csc 10.	*l* tan 9.	d 1′	*l* cot 10.	*l* sec 10.	d 1′	*l* cos 9.	′
0	31788	59	68212	32747	63	67253	00960	2	99040	60
1	847	60	153	810	62	190	962	3	038	59
2	907	59	093	872	61	128	965	3	035	58
3	966	59	034	933	62	067	968	2	032	57
4	32025	59	67975	995	62	005	970	3	030	56
5	084	59	916	33057	62	66943	973	3	027	55
6	143	59	857	119	61	881	976	2	024	54
7	202	59	798	180	62	820	978	3	022	53
8	261	58	739	242	61	758	981	3	019	52
9	319	59	681	303	62	697	984	3	016	51
10	378	59	622	365	61	635	987	2	013	50
11	437	58	563	426	61	574	989	3	011	49
12	495	58	505	487	61	513	992	3	008	48
13	553	59	447	548	61	452	995	3	005	47
14	612	58	388	609	61	391	998	2	002	46
15	670	58	330	670	61	330	01000	3	000	45
16	728	58	272	731	61	269	003	3	98997	44
17	786	58	214	792	61	208	006	3	994	43
18	844	58	156	853	60	147	009	2	991	42
19	902	58	098	913	61	087	011	3	989	41
20	960	58	040	974	60	026	014	3	986	40
21	33018	57	66982	34034	61	65966	017	3	983	39
22	075	58	925	095	60	905	020	2	980	38
23	133	57	867	155	60	845	022	3	978	37
24	190	58	810	215	61	785	025	3	975	36
25	248	57	752	276	60	724	028	3	972	35
26	305	57	695	336	60	664	031	2	969	34
27	362	58	638	396	60	604	033	3	967	33
28	420	57	580	456	60	544	036	3	964	32
29	477	57	523	516	60	484	039	3	961	31
30	33534	57	66466	34576	59	65424	01042	3	98958	30
31	591	56	409	635	60	365	045	2	955	29
32	647	57	353	695	60	305	047	3	953	28
33	704	57	296	755	59	245	050	3	950	27
34	761	57	239	814	60	186	053	3	947	26
35	818	56	182	874	59	126	056	3	944	25
36	874	57	126	933	59	067	059	3	941	24
37	931	56	069	992	59	008	062	2	938	23
38	987	56	013	35051	60	64949	064	3	936	22
39	34043	57	65957	111	59	889	067	3	933	21
40	100	56	900	170	59	830	070	3	930	20
41	156	56	844	229	59	771	073	3	927	19
42	212	56	788	288	59	712	076	3	924	18
43	268	56	732	347	58	653	079	2	921	17
44	324	56	676	405	59	595	081	3	919	16
45	380	56	620	464	59	536	084	3	916	15
46	436	55	564	523	58	477	087	3	913	14
47	491	56	509	581	59	419	090	3	910	13
48	547	55	453	640	58	360	093	3	907	12
49	602	56	398	698	59	302	096	3	904	11
50	658	55	342	757	58	243	099	3	901	10
51	713	56	287	815	58	185	102	2	898	9
52	769	55	231	873	58	127	104	3	896	8
53	824	55	176	931	58	069	107	3	893	7
54	879	55	121	989	58	011	110	3	890	6
55	934	55	066	36047	58	63953	113	3	887	5
56	989	55	011	105	58	895	116	3	884	4
57	35044	55	64956	163	58	837	119	3	881	3
58	099	55	901	221	58	779	122	3	878	2
59	154	55	846	279	57	721	125	3	875	1
60	35209		64791	36336		63664	01128		98872	0
′	9. *l* cos	d 1′	10. *l* sec	9. *l* cot	d 1′	10. *l* tan	10. *l* csc	d 1′	9. *l* sin	′

Proportional Parts

″	63	62	61	60	59	58	57	56	55	3	2
0	0	0	0	0	0	0	0	0	0	0	0
1	1	1	1	1	1	1	1	1	1	0	0
2	2	2	2	2	2	2	2	2	2	0	0
3	3	3	3	3	3	3	3	3	3	0	0
4	4	4	4	4	4	4	4	4	4	0	0
5	5	5	5	5	5	5	5	5	5	0	0
6	6	6	6	6	6	6	6	6	6	0	0
7	7	7	7	7	7	7	7	7	6	0	0
8	8	8	8	8	8	8	8	7	7	0	0
9	9	9	9	9	9	9	9	8	8	0	0
10	10	10	10	10	10	10	10	9	9	0	0
11	12	11	11	11	11	11	10	10	10	1	0
12	13	12	12	12	12	12	11	11	11	1	0
13	14	13	13	13	13	13	12	12	12	1	0
14	15	14	14	14	14	14	13	13	13	1	0
15	16	15	15	15	15	14	14	14	14	1	0
16	17	17	16	16	16	15	15	15	15	1	1
17	18	18	17	17	17	16	16	16	16	1	1
18	19	19	18	18	18	17	17	17	16	1	1
19	20	20	19	19	19	18	18	18	17	1	1
20	21	21	20	20	20	19	19	19	18	1	1
21	22	22	21	21	21	20	20	20	19	1	1
22	23	23	22	22	22	21	21	21	20	1	1
23	24	24	23	23	23	22	22	21	21	1	1
24	25	25	24	24	24	23	23	22	22	1	1
25	26	26	25	25	25	24	24	23	23	1	1
26	27	27	26	26	26	25	25	24	24	1	1
27	28	28	27	27	27	26	26	25	25	1	1
28	29	29	28	28	28	27	27	26	26	1	1
29	30	30	29	29	29	28	28	27	27	1	1
30	32	31	30	30	30	29	28	28	28	2	1
31	33	32	32	31	30	30	29	29	28	2	1
32	34	33	33	32	31	31	30	30	29	2	1
33	35	34	34	33	32	32	31	31	30	2	1
34	36	35	35	34	33	33	32	32	31	2	1
35	37	36	36	35	34	34	33	33	32	2	1
36	38	37	37	36	35	35	34	34	33	2	1
37	39	38	38	37	36	36	35	35	34	2	1
38	40	39	39	38	37	37	36	35	35	2	1
39	41	40	40	39	38	38	37	36	36	2	1
40	42	41	41	40	39	39	38	37	37	2	1
41	43	42	42	41	40	40	39	38	38	2	1
42	44	43	43	42	41	41	40	39	38	2	1
43	45	44	44	43	42	42	41	40	39	2	1
44	46	45	45	44	43	43	42	41	40	2	1
45	47	47	46	45	44	44	43	42	41	2	2
46	48	48	47	46	45	44	44	43	42	2	2
47	49	49	48	47	46	45	45	44	43	2	2
48	50	50	49	48	47	46	46	45	44	2	2
49	51	51	50	49	48	47	47	46	45	2	2
50	52	52	51	50	49	48	48	47	46	2	2
51	54	53	52	51	50	49	48	48	47	3	2
52	55	54	53	52	51	50	49	49	48	3	2
53	56	55	54	53	52	51	50	49	49	3	2
54	57	56	55	54	53	52	51	50	50	3	2
55	58	57	56	55	54	53	52	51	50	3	2
56	59	58	57	56	55	54	53	52	51	3	2
57	60	59	58	57	56	55	54	53	52	3	2
58	61	60	59	58	57	56	55	54	53	3	2
59	62	61	60	59	58	57	56	55	54	3	2
60	63	62	61	60	59	58	57	56	55	3	2
″	63	62	61	60	59	58	57	56	55	3	2

Proportional Parts

′	l sin 9.	d 1′	l csc 10.	l tan 9.	d 1′	l cot 10.	l sec 10.	d 1′	l cos 9.	′
0	35209	54	64791	36336	58	63664	01128	3	98872	60
1	263	55	737	394	58	606	131	2	869	59
2	318	55	682	452	57	548	133	3	867	58
3	373	54	627	509	57	491	136	3	864	57
4	427	54	573	566	58	434	139	3	861	56
5	481	55	519	624	57	376	142	3	858	55
6	536	54	464	681	57	319	145	3	855	54
7	590	54	410	738	57	262	148	3	852	53
8	644	54	356	795	57	205	151	3	849	52
9	698	54	302	852	57	148	154	3	846	51
10	752	54	248	909	57	091	157	3	843	50
11	806	54	194	966	57	034	160	3	840	49
12	860	54	140	37023	57	62977	163	3	837	48
13	914	54	086	080	57	920	166	3	834	47
14	968	54	032	137	56	863	169	3	831	46
15	36022	53	63978	193	57	807	172	3	828	45
16	075	54	925	250	56	750	175	3	825	44
17	129	53	871	306	57	694	178	3	822	43
18	182	54	818	363	56	637	181	3	819	42
19	236	53	764	419	57	581	184	3	816	41
20	289	53	711	476	56	524	187	3	813	40
21	342	53	658	532	56	468	190	3	810	39
22	395	54	605	588	56	412	193	3	807	38
23	449	53	551	644	56	356	196	3	804	37
24	502	53	498	700	56	300	199	3	801	36
25	555	53	445	756	56	244	202	3	798	35
26	608	52	392	812	56	188	205	3	795	34
27	660	53	340	868	56	132	208	3	792	33
28	713	53	287	924	56	076	211	3	789	32
29	766	53	234	980	55	020	214	3	786	31
30	36819	52	63181	38035	56	61965	01217	3	98783	30
31	871	53	129	091	56	909	220	3	780	29
32	924	52	076	147	55	853	223	3	777	28
33	976	52	024	202	55	798	226	3	774	27
34	37028	53	62972	257	56	743	229	3	771	26
35	081	52	919	313	55	687	232	3	768	25
36	133	52	867	368	55	632	235	3	765	24
37	185	52	815	423	56	577	238	3	762	23
38	237	52	763	479	55	521	241	3	759	22
39	289	52	711	534	55	466	244	3	756	21
40	341	52	659	589	55	411	247	3	753	20
41	393	52	607	644	55	356	250	4	750	19
42	445	52	555	699	55	301	254	3	746	18
43	497	52	503	754	54	246	257	3	743	17
44	549	51	451	808	55	192	260	3	740	16
45	600	52	400	863	55	137	263	3	737	15
46	652	51	348	918	54	082	266	3	734	14
47	703	52	297	972	55	028	269	3	731	13
48	755	51	245	39027	55	60973	272	3	728	12
49	806	52	194	082	54	918	275	3	725	11
50	858	51	142	136	54	864	278	3	722	10
51	909	51	091	190	55	810	281	4	719	9
52	960	51	040	245	54	755	285	3	715	8
53	38011	51	61989	299	54	701	288	3	712	7
54	062	51	938	353	54	647	291	3	709	6
55	113	51	887	407	54	593	294	3	706	5
56	164	51	836	461	54	539	297	3	703	4
57	215	51	785	515	54	485	300	3	700	3
58	266	51	734	569	54	431	303	3	697	2
59	317	51	683	623	54	377	306	4	694	1
60	38368		61632	39677		60323	01310		98690	0
′	9. l cos	d 1′	10. l sec	9. l cot	d 1′	10. l tan	10. l csc	d 1′	9. l sin	′

Proportional Parts

″	58	57	56	55	54	53	52	51	4	3	2
0	0	0	0	0	0	0	0	0	0	0	0
1	1	1	1	1	1	1	1	1	0	0	0
2	2	2	2	2	2	2	2	2	0	0	0
3	3	3	3	3	3	3	3	3	0	0	0
4	4	4	4	4	4	4	3	3	0	0	0
5	5	5	5	5	4	4	4	4	0	0	0
6	6	6	6	6	5	5	5	5	0	0	0
7	7	7	7	6	6	6	6	6	0	0	0
8	8	8	7	7	7	7	7	7	1	0	0
9	9	9	8	8	8	8	8	8	1	0	0
10	10	10	9	9	9	9	9	8	1	0	0
11	11	10	10	10	10	10	10	9	1	1	0
12	12	11	11	11	11	11	10	10	1	1	0
13	13	12	12	12	12	11	11	11	1	1	0
14	14	13	13	13	13	12	12	12	1	1	0
15	14	14	14	14	14	13	13	13	1	1	0
16	15	15	15	15	14	14	14	14	1	1	1
17	16	16	16	16	15	15	15	14	1	1	1
18	17	17	17	16	16	16	16	15	1	1	1
19	18	18	18	17	17	17	16	16	1	1	1
20	19	19	19	18	18	18	17	17	1	1	1
21	20	20	20	19	19	19	18	18	1	1	1
22	21	21	21	20	20	19	19	19	1	1	1
23	22	22	21	21	21	20	20	20	2	1	1
24	23	23	22	22	22	21	21	20	2	1	1
25	24	24	23	23	22	22	22	21	2	1	1
26	25	25	24	24	23	23	23	22	2	1	1
27	26	26	25	25	24	24	23	23	2	1	1
28	27	27	26	26	25	25	24	24	2	1	1
29	28	28	27	27	26	26	25	25	2	1	1
30	29	28	28	28	27	26	26	26	2	2	1
31	30	29	29	28	28	27	27	26	2	2	1
32	31	30	30	29	29	28	28	27	2	2	1
33	32	31	31	30	30	29	29	28	2	2	1
34	33	32	32	31	31	30	29	29	2	2	1
35	34	33	33	32	32	31	30	30	2	2	1
36	35	34	34	33	32	32	31	31	2	2	1
37	36	35	35	34	33	33	32	31	2	2	1
38	37	36	35	35	34	34	33	32	3	2	1
39	38	37	36	36	35	34	34	33	3	2	1
40	39	38	37	37	36	35	35	34	3	2	1
41	40	39	38	38	37	36	36	35	3	2	1
42	41	40	39	38	38	37	36	36	3	2	1
43	42	41	40	39	39	38	37	37	3	2	1
44	43	42	41	40	40	39	38	37	3	2	1
45	44	43	42	41	40	40	39	38	3	2	2
46	44	44	43	42	41	41	40	39	3	2	2
47	45	45	44	43	42	42	41	40	3	2	2
48	46	46	45	44	43	42	42	41	3	2	2
49	47	47	46	45	44	43	42	42	3	2	2
50	48	48	47	46	45	44	43	42	3	2	2
51	49	48	48	47	46	45	44	43	3	3	2
52	50	49	49	48	47	46	45	44	3	3	2
53	51	50	49	49	48	47	46	45	4	3	2
54	52	51	50	50	49	48	47	46	4	3	2
55	53	52	51	50	50	49	48	47	4	3	2
56	54	53	52	51	50	49	49	48	4	3	2
57	55	54	53	52	51	50	49	48	4	3	2
58	56	55	54	53	52	51	50	49	4	3	2
59	57	56	55	54	53	52	51	50	4	3	2
60	58	57	56	55	54	53	52	51	4	3	2
″	58	57	56	55	54	53	52	51	4	3	2

Proportional Parts

14° TABLE II **165°**

′	l sin 9.	d 1′	l csc 10.	l tan 9.	d 1′	l cot 10.	l sec 10.	d 1′	l cos 9.	′
0	38368		61632	39677		60323	01310		98690	60
1	418	50	582	731	54	269	313	3	687	59
2	469	51	531	785	54	215	316	3	684	58
3	519	50	481	838	53	162	319	3	681	57
4	570	51	430	892	54	108	322	3	678	56
5	620	50	380	945	53	055	325	3	675	55
6	670	50	330	999	54	001	329	4	671	54
7	721	51	279	40052	53	59948	332	3	668	53
8	771	50	229	106	54	894	335	3	665	52
9	821	50	179	159	53	841	338	3	662	51
10	871	50	129	212	53	788	341	3	659	50
11	921	50	079	266	54	734	344	3	656	49
12	971	50	029	319	53	681	348	4	652	48
13	39021	50	60979	372	53	628	351	3	649	47
14	071	50	929	425	53	575	354	3	646	46
15	121	50	879	478	53	522	357	3	643	45
16	170	49	830	531	53	469	360	3	640	44
17	220	50	780	584	53	416	364	4	636	43
18	270	50	730	636	52	364	367	3	633	42
19	319	49	681	689	53	311	370	3	630	41
20	369	50	631	742	53	258	373	3	627	40
21	418	49	582	795	53	205	377	4	623	39
22	467	49	533	847	52	153	380	3	620	38
23	517	50	483	900	53	100	383	3	617	37
24	566	49	434	952	52	048	386	3	614	36
25	615	49	385	41005	53	58995	390	4	610	35
26	664	49	336	057	52	943	393	3	607	34
27	713	49	287	109	52	891	396	3	604	33
28	762	49	238	161	52	839	399	3	601	32
29	811	49	189	214	53	786	403	4	597	31
30	39860	49	60140	41266	52	58734	01406	3	98594	30
31	909	49	091	318	52	682	409	3	591	29
32	958	49	042	370	52	630	412	3	588	28
33	40006	48	59994	422	52	578	416	4	584	27
34	055	49	945	474	52	526	419	3	581	26
35	103	48	897	526	52	474	422	3	578	25
36	152	49	848	578	52	422	426	4	574	24
37	200	48	800	629	51	371	429	3	571	23
38	249	49	751	681	52	319	432	3	568	22
39	297	48	703	733	52	267	435	3	565	21
40	346	49	654	784	51	216	439	4	561	20
41	394	48	606	836	52	164	442	3	558	19
42	442	48	558	887	51	113	445	3	555	18
43	490	48	510	939	52	061	449	4	551	17
44	538	48	462	990	51	010	452	3	548	16
45	586	48	414	42041	51	57959	455	3	545	15
46	634	48	366	093	52	907	459	4	541	14
47	682	48	318	144	51	856	462	3	538	13
48	730	48	270	195	51	805	465	3	535	12
49	778	48	222	246	51	754	469	4	531	11
50	825	47	175	297	51	703	472	3	528	10
51	873	48	127	348	51	652	475	3	525	9
52	921	48	079	399	51	601	479	4	521	8
53	968	47	032	450	51	550	482	3	518	7
54	41016	48	58984	501	51	499	485	3	515	6
55	063	47	937	552	51	448	489	4	511	5
56	111	48	889	603	51	397	492	3	508	4
57	158	47	842	653	50	347	495	3	505	3
58	205	47	795	704	51	296	499	4	501	2
59	252	47	748	755	51	245	502	3	498	1
60	41300	48	58700	42805	50	57195	01506	4	98494	0
′	9. l cos	d 1′	10. l sec	9. l cot	d 1′	10. l tan	10. l csc	d 1′	9. l sin	′

Proportional Parts

″	54	53	52	51	50	49	48	47	4	3
0	0	0	0	0	0	0	0	0	0	0
1	1	1	1	1	1	1	1	1	0	0
2	2	2	2	2	2	2	2	2	0	0
3	3	3	3	3	2	2	2	2	0	0
4	4	4	3	3	3	3	3	3	0	0
5	4	4	4	4	4	4	4	4	0	0
6	5	5	5	5	5	5	5	5	0	0
7	6	6	6	6	6	6	6	5	0	0
8	7	7	7	7	7	7	6	6	1	0
9	8	8	8	8	8	7	7	7	1	0
10	9	9	9	8	8	8	8	8	1	0
11	10	10	10	9	9	9	9	9	1	1
12	11	11	10	10	10	10	10	9	1	1
13	12	11	11	11	11	11	10	10	1	1
14	13	12	12	12	12	11	11	11	1	1
15	14	13	13	13	12	12	12	12	1	1
16	14	14	14	14	13	13	13	13	1	1
17	15	15	15	14	14	14	14	13	1	1
18	16	16	16	15	15	15	14	14	1	1
19	17	17	16	16	16	16	15	15	1	1
20	18	18	17	17	17	16	16	16	1	1
21	19	19	18	18	18	17	17	16	1	1
22	20	19	19	19	18	18	18	17	1	1
23	21	20	20	20	19	19	18	18	2	1
24	22	21	21	20	20	20	19	19	2	1
25	22	22	22	21	21	20	20	20	2	1
26	23	23	23	22	22	21	21	20	2	1
27	24	24	23	23	22	22	22	21	2	1
28	25	25	24	24	23	23	22	22	2	1
29	26	26	25	25	24	24	23	23	2	1
30	27	26	26	26	25	24	24	24	2	2
31	28	27	27	26	26	25	25	24	2	2
32	29	28	28	27	27	26	26	25	2	2
33	30	29	29	28	28	27	26	26	2	2
34	31	30	29	29	28	28	27	27	2	2
35	32	31	30	30	29	29	28	27	2	2
36	32	32	31	31	30	29	29	28	2	2
37	33	33	32	31	31	30	30	29	2	2
38	34	34	33	32	32	31	30	30	3	2
39	35	34	34	33	33	32	31	31	3	2
40	36	35	35	34	33	33	32	31	3	2
41	37	36	36	35	34	33	33	32	3	2
42	38	37	36	36	35	34	34	33	3	2
43	39	38	37	37	36	35	34	34	3	2
44	40	39	38	37	37	36	35	34	3	2
45	40	40	39	38	38	37	36	35	3	2
46	41	41	40	39	38	38	37	36	3	2
47	42	42	41	40	39	38	38	37	3	2
48	43	42	42	41	40	39	38	38	3	2
49	44	43	42	42	41	40	39	38	3	2
50	45	44	43	42	42	41	40	39	3	2
51	46	45	44	43	42	42	41	40	3	3
52	47	46	45	44	43	42	42	41	3	3
53	48	47	46	45	44	43	42	42	4	3
54	49	48	47	46	45	44	43	42	4	3
55	50	49	48	47	46	45	44	43	4	3
56	50	49	49	48	47	46	45	44	4	3
57	51	50	49	48	48	47	46	45	4	3
58	52	51	50	49	48	47	46	45	4	3
59	53	52	51	50	49	48	47	46	4	3
60	54	53	52	51	50	49	48	47	4	3
″	54	53	52	51	50	49	48	47	4	3

Proportional Parts

104° **75°**

′	l sin 9.	d 1′	l csc 10.	l tan 9.	d 1′	l cot 10.	l sec 10.	d 1′	l cos 9.	′
0	41300	47	58700	42805	51	57195	01506	3	98494	60
1	347	47	653	856	50	144	509	3	491	59
2	394	47	606	906	51	094	512	4	488	58
3	441	47	559	957	50	043	516	3	484	57
4	488	47	512	43007	50	56993	519	4	481	56
5	535	47	465	057	51	943	523	3	477	55
6	582	46	418	108	50	892	526	3	474	54
7	628	47	372	158	50	842	529	4	471	53
8	675	47	325	208	50	792	533	3	467	52
9	722	46	278	258	50	742	536	4	464	51
10	768	47	232	308	50	692	540	3	460	50
11	815	46	185	358	50	642	543	4	457	49
12	861	47	139	408	50	592	547	3	453	48
13	908	46	092	458	50	542	550	3	450	47
14	954	47	046	508	50	492	553	4	447	46
15	42001	46	57999	558	49	442	557	3	443	45
16	047	46	953	607	50	393	560	4	440	44
17	093	47	907	657	50	343	564	3	436	43
18	140	46	860	707	49	293	567	4	433	42
19	186	46	814	756	50	244	571	3	429	41
20	232	46	768	806	49	194	574	4	426	40
21	278	46	722	855	50	145	578	3	422	39
22	324	46	676	905	49	095	581	4	419	38
23	370	46	630	954	50	046	585	3	415	37
24	416	45	584	44004	49	55996	588	3	412	36
25	461	46	539	053	49	947	591	4	409	35
26	507	46	493	102	49	898	595	3	405	34
27	553	46	447	151	50	849	598	4	402	33
28	599	45	401	201	49	799	602	3	398	32
29	644	46	356	250	49	750	605	4	395	31
30	42690	45	57310	44299	49	55701	01609	3	98391	30
31	735	46	265	348	49	652	612	4	388	29
32	781	45	219	397	49	603	616	3	384	28
33	826	46	174	446	49	554	619	4	381	27
34	872	45	128	495	49	505	623	4	377	26
35	917	45	083	544	48	456	627	3	373	25
36	962	46	038	592	49	408	630	4	370	24
37	43008	45	56992	641	49	359	634	3	366	23
38	053	45	947	690	48	310	637	4	363	22
39	098	45	902	738	49	262	641	3	359	21
40	143	45	857	787	49	213	644	4	356	20
41	188	45	812	836	48	164	648	3	352	19
42	233	45	767	884	49	116	651	4	349	18
43	278	45	722	933	48	067	655	3	345	17
44	323	44	677	981	48	019	658	4	342	16
45	367	45	633	45029	49	54971	662	4	338	15
46	412	45	588	078	48	922	666	3	334	14
47	457	45	543	126	48	874	669	4	331	13
48	502	44	498	174	48	826	673	3	327	12
49	546	45	454	222	49	778	676	4	324	11
50	591	44	409	271	48	729	680	3	320	10
51	635	45	365	319	48	681	683	4	317	9
52	680	44	320	367	48	633	687	4	313	8
53	724	45	276	415	48	585	691	3	309	7
54	769	44	231	463	48	537	694	4	306	6
55	813	44	187	511	48	489	698	3	302	5
56	857	44	143	559	47	441	701	4	299	4
57	901	45	099	606	48	394	705	4	295	3
58	946	44	054	654	48	346	709	3	291	2
59	990	44	010	702	48	298	712	4	288	1
60	44034		55966	45750		54250	01716		98284	0
′	9. l cos	d 1′	10. l sec	9. l cot	d 1′	10. l tan	10. l csc	d 1′	9. l sin	′

105° 74°

Proportional Parts

″	51	50	49	48	47	46	45	44	4	3
0	0	0	0	0	0	0	0	0	0	0
1	1	1	1	1	1	1	1	1	0	0
2	2	2	2	2	2	2	2	1	0	0
3	3	2	2	2	2	2	2	2	0	0
4	3	3	3	3	3	3	3	3	0	0
5	4	4	4	4	4	4	4	4	0	0
6	5	5	5	5	5	5	4	4	0	0
7	6	6	6	6	5	5	5	5	0	0
8	7	7	7	6	6	6	6	6	1	0
9	8	8	7	7	7	7	7	7	1	0
10	8	8	8	8	8	8	8	7	1	0
11	9	9	9	9	9	8	8	8	1	1
12	10	10	10	10	9	9	9	9	1	1
13	11	11	11	10	10	10	10	10	1	1
14	12	12	11	11	11	11	10	10	1	1
15	13	12	12	12	12	12	11	11	1	1
16	14	13	13	13	13	12	12	12	1	1
17	14	14	14	14	13	13	13	12	1	1
18	15	15	15	14	14	14	14	13	1	1
19	16	16	16	15	15	15	14	14	1	1
20	17	17	16	16	16	15	15	15	1	1
21	18	18	17	17	16	16	16	15	1	1
22	19	18	18	18	17	17	16	16	1	1
23	20	19	19	18	18	18	17	17	2	1
24	20	20	20	19	19	18	18	18	2	1
25	21	21	20	20	20	19	19	18	2	1
26	22	22	21	21	20	20	20	19	2	1
27	23	22	22	22	21	21	20	20	2	1
28	24	23	23	22	22	21	21	21	2	1
29	25	24	24	23	23	22	22	21	2	1
30	26	25	24	24	24	23	22	22	2	2
31	26	26	25	25	24	24	23	23	2	2
32	27	27	26	26	25	25	24	23	2	2
33	28	28	27	26	26	25	25	24	2	2
34	29	28	28	27	27	26	26	25	2	2
35	30	29	29	28	27	27	26	26	2	2
36	31	30	29	29	28	28	27	26	2	2
37	31	31	30	30	29	28	28	27	2	2
38	32	32	31	30	30	29	28	28	3	2
39	33	33	32	31	31	30	29	29	3	2
40	34	33	33	32	31	31	30	29	3	2
41	35	34	33	33	32	31	31	30	3	2
42	36	35	34	34	33	32	32	31	3	2
43	37	36	35	34	34	33	32	32	3	2
44	37	37	36	35	34	34	33	32	3	2
45	38	38	37	36	35	34	34	33	3	2
46	39	38	38	37	36	35	34	34	3	2
47	40	39	38	38	37	36	35	34	3	2
48	41	40	39	38	38	37	36	35	3	2
49	42	41	40	39	38	38	37	36	3	2
50	42	42	41	40	39	38	38	37	3	2
51	43	42	42	41	40	39	38	37	3	3
52	44	43	42	42	41	40	39	38	3	3
53	45	44	43	42	42	41	40	39	4	3
54	46	45	44	43	42	41	40	40	4	3
55	47	46	45	44	43	42	41	40	4	3
56	48	47	46	45	44	43	42	41	4	3
57	48	48	47	46	45	44	43	42	4	3
58	49	48	47	46	45	44	44	43	4	3
59	50	49	48	47	46	45	44	43	4	3
60	51	50	49	48	47	46	45	44	4	3
″	51	50	49	48	47	46	45	44	4	3

Proportional Parts

′	*l* sin 9.	d 1′	*l* csc 10.	*l* tan 9.	d 1′	*l* cot 10.	*l* sec 10.	d 1′	*l* cos 9.	′
0	**44034**	44	**55966**	**45750**	47	**54250**	**01716**	3	**98284**	**60**
1	078	44	922	797	48	203	719	4	281	59
2	122	44	878	845	47	155	723	4	277	58
3	166	44	834	892	48	108	727	3	273	57
4	210	43	790	940	47	060	730	4	270	56
5	253	44	747	987	48	013	734	4	266	**55**
6	297	44	703	**46035**	47	**53965**	738	3	262	54
7	341	44	659	082	48	918	741	4	259	53
8	385	43	615	130	47	870	745	4	255	52
9	428	44	572	177	47	823	749	3	251	51
10	472	44	528	224	47	776	752	4	248	**50**
11	516	43	484	271	48	729	756	4	244	49
12	559	43	441	319	47	681	760	3	240	48
13	602	44	398	366	47	634	763	4	237	47
14	646	43	354	413	47	587	767	4	233	46
15	689	44	311	460	47	540	771	3	229	**45**
16	733	43	267	507	47	493	774	4	226	44
17	776	43	224	554	47	446	778	4	222	43
18	819	43	181	601	47	399	782	3	218	42
19	862	43	138	648	46	352	785	4	215	41
20	905	43	095	694	47	306	789	4	211	**40**
21	948	44	052	741	47	259	793	3	207	39
22	992	43	008	788	47	212	796	4	204	38
23	**45035**	42	**54965**	835	46	165	800	4	200	37
24	077	43	923	881	47	119	804	4	196	36
25	120	43	880	928	47	072	808	3	192	**35**
26	163	43	837	975	46	025	811	4	189	34
27	206	43	794	**47021**	47	**52979**	815	4	185	33
28	249	43	751	068	46	932	819	4	181	32
29	292	42	708	114	46	886	823	3	177	31
30	**45334**	43	**54666**	**47160**	47	**52840**	**01826**	4	**98174**	**30**
31	377	42	623	207	46	793	830	4	170	29
32	419	43	581	253	46	747	834	4	166	28
33	462	42	538	299	47	701	838	3	162	27
34	504	43	496	346	46	654	841	4	159	26
35	547	42	453	392	46	608	845	4	155	**25**
36	589	43	411	438	46	562	849	4	151	24
37	632	42	368	484	46	516	853	3	147	23
38	674	42	326	530	46	470	856	4	144	22
39	716	42	284	576	46	424	860	4	140	21
40	758	43	242	622	46	378	864	4	136	**20**
41	801	42	199	668	46	332	868	3	132	19
42	843	42	157	714	46	286	871	4	129	18
43	885	42	115	760	46	240	875	4	125	17
44	927	42	073	806	46	194	879	4	121	16
45	969	42	031	852	45	148	883	4	117	**15**
46	**46011**	42	**53989**	897	46	103	887	3	113	14
47	053	42	947	943	46	057	890	4	110	13
48	095	41	905	989	46	011	894	4	106	12
49	136	42	864	**48035**	45	**51965**	898	4	102	11
50	178	42	822	080	46	920	902	4	098	**10**
51	220	42	780	126	45	874	906	4	094	9
52	262	41	738	171	46	829	910	3	090	8
53	303	42	697	217	45	783	913	4	087	7
54	345	41	655	262	45	738	917	4	083	6
55	386	42	614	307	46	693	921	4	079	**5**
56	428	41	572	353	45	647	925	4	075	4
57	469	42	531	398	45	602	929	4	071	3
58	511	41	489	443	46	557	933	4	067	2
59	552	42	448	489	45	511	937	3	063	1
60	**46594**		**53406**	**48534**		**51466**	**01940**		**98060**	**0**
′	9. *l* cos	d 1′	10. *l* sec	9. *l* cot	d 1′	10. *l* tan	10. *l* csc	d 1′	9. *l* sin	′

Proportional Parts

″	48	47	46	45	44	43	42	41	4	3
0	0	0	0	0	0	0	0	0	0	0
1	1	1	1	1	**1**	**1**	**1**	**1**	0	0
2	**2**	2	2	2	1	1	1	1	0	0
3	2	**2**	**2**	**2**	2	2	2	2	0	0
4	3	3	3	3	3	3	3	**3**	0	0
5	4	4	4	**4**	**4**	4	4	3	0	0
6	5	**5**	5	4	4	**4**	**4**	4	0	0
7	6	5	**5**	5	5	5	5	**5**	0	0
8	**6**	6	6	6	6	**6**	6	5	1	0
9	7	7	7	7	7	6	**6**	6	1	0
10	8	8	**8**	8	**7**	7	7	7	1	0
11	9	**9**	8	**8**	8	8	**8**	8	1	1
12	**10**	9	9	9	9	9	8	**8**	1	1
13	10	10	10	**10**	10	**9**	9	9	1	1
14	11	11	11	10	**10**	10	**10**	10	1	1
15	12	12	12	11	11	**11**	10	**10**	**1**	1
16	13	13	**12**	12	**12**	11	11	11	1	1
17	14	**13**	13	13	12	12	12	12	1	1
18	**14**	14	14	14	13	13	13	**12**	1	1
19	15	15	15	**14**	14	14	**13**	13	1	1
20	16	**16**	**15**	15	**15**	**14**	14	**14**	1	**1**
21	17	16	16	**16**	15	15	**15**	14	1	1
22	**18**	17	17	16	16	**16**	15	15	1	1
23	18	18	**18**	17	17	16	16	**16**	2	1
24	19	19	18	18	18	17	17	16	2	1
25	20	20	19	19	**18**	18	18	17	2	1
26	21	**20**	20	20	19	19	**18**	**18**	2	1
27	22	21	**21**	**20**	20	**19**	19	18	2	1
28	**22**	22	21	21	21	20	20	19	2	1
29	23	23	22	**22**	**21**	21	**20**	**20**	2	1
30	24	24	23	22	22	22	21	20	**2**	2
31	25	**24**	24	23	**23**	**22**	**22**	21	2	2
32	**26**	25	25	24	23	23	22	22	2	2
33	26	26	**25**	25	24	**24**	23	23	2	2
34	27	**27**	26	26	25	24	**24**	**23**	2	2
35	28	27	27	**26**	**26**	25	24	24	2	2
36	29	28	28	27	26	26	25	25	2	2
37	30	29	**28**	**28**	27	27	26	**25**	2	2
38	**30**	30	29	28	28	**27**	27	26	3	2
39	31	31	30	29	29	28	**27**	27	3	2
40	32	**31**	**31**	30	**29**	**29**	28	**27**	3	**2**
41	33	32	31	31	30	29	**29**	28	3	2
42	**34**	33	32	32	31	30	29	**29**	3	2
43	34	**34**	33	**32**	32	31	30	29	3	2
44	35	34	**34**	33	**32**	32	31	30	3	2
45	36	35	34	**34**	33	**32**	32	**31**	**3**	2
46	37	36	35	34	**34**	33	**32**	31	3	2
47	38	37	36	35	34	**34**	33	32	3	2
48	**38**	38	37	36	35	34	34	**33**	3	2
49	39	**38**	38	37	36	35	**34**	33	3	2
50	40	39	**38**	38	**37**	36	35	34	3	2
51	41	40	39	**38**	37	37	**36**	35	3	3
52	**42**	41	40	39	38	**37**	36	36	3	3
53	42	42	**41**	**40**	39	38	37	**36**	4	3
54	43	**42**	41	40	40	**39**	**38**	37	4	3
55	44	43	42	41	**40**	39	38	38	4	3
56	45	44	43	42	41	40	39	**38**	4	3
57	46	**45**	**44**	43	42	41	40	39	4	3
58	**46**	45	44	44	43	42	41	40	4	3
59	47	46	45	**44**	**43**	**42**	**41**	**40**	4	3
60	48	47	46	45	44	43	42	41	**4**	**3**
″	48	47	46	45	44	43	42	41	4	3

Proportional Parts

′	l sin 9.	d 1′	l csc 10.	l tan 9.	d 1′	l cot 10.	l sec 10.	d 1′	l cos 9.	′
0	46594	41	53406	48534	45	51466	01940	4	98060	60
1	635	41	365	579	45	421	944	4	056	59
2	676	41	324	624	45	376	948	4	052	58
3	717	41	283	669	45	331	952	4	048	57
4	758	42	242	714	45	286	956	4	044	56
5	800	41	200	759	45	241	960	4	040	55
6	841	41	159	804	45	196	964	4	036	54
7	882	41	118	849	45	151	968	3	032	53
8	923	41	077	894	45	106	971	4	029	52
9	964	41	036	939	45	061	975	4	025	51
10	47005	40	52995	984	45	016	979	4	021	50
11	045	41	955	49029	44	50971	983	4	017	49
12	086	41	914	073	45	927	987	4	013	48
13	127	41	873	118	45	882	991	4	009	47
14	168	41	832	163	44	837	995	4	005	46
15	209	40	791	207	45	793	999	4	001	45
16	249	41	751	252	44	748	02003	4	97997	44
17	290	40	710	296	45	704	007	4	993	43
18	330	41	670	341	44	659	011	3	989	42
19	371	40	629	385	45	615	014	4	986	41
20	411	41	589	430	44	570	018	4	982	40
21	452	40	548	474	45	526	022	4	978	39
22	492	41	508	519	44	481	026	4	974	38
23	533	40	467	563	44	437	030	4	970	37
24	573	40	427	607	45	393	034	4	966	36
25	613	41	387	652	44	348	038	4	962	35
26	654	40	346	696	44	304	042	4	958	34
27	694	40	306	740	44	260	046	4	954	33
28	734	40	266	784	44	216	050	4	950	32
29	774	40	226	828	44	172	054	4	946	31
30	47814	40	52186	49872	44	50128	02058	4	97942	30
31	854	40	146	916	44	084	062	4	938	29
32	894	40	106	960	44	040	066	4	934	28
33	934	40	066	50004	44	49996	070	4	930	27
34	974	40	026	048	44	952	074	4	926	26
35	48014	40	51986	092	44	908	078	4	922	25
36	054	40	946	136	44	864	082	4	918	24
37	094	39	906	180	43	820	086	4	914	23
38	133	40	867	223	44	777	090	4	910	22
39	173	40	827	267	44	733	094	4	906	21
40	213	39	787	311	44	689	098	4	902	20
41	252	40	748	355	43	645	102	4	898	19
42	292	40	708	398	44	602	106	4	894	18
43	332	39	668	442	43	558	110	4	890	17
44	371	40	629	485	44	515	114	4	886	16
45	411	39	589	529	43	471	118	4	882	15
46	450	40	550	572	44	428	122	4	878	14
47	490	39	510	616	43	384	126	4	874	13
48	529	39	471	659	44	341	130	4	870	12
49	568	39	432	703	43	297	134	5	866	11
50	607	40	393	746	43	254	139	4	861	10
51	647	39	353	789	44	211	143	4	857	9
52	686	39	314	833	43	167	147	4	853	8
53	725	39	275	876	43	124	151	4	849	7
54	764	39	236	919	43	081	155	4	845	6
55	803	39	197	962	43	038	159	4	841	5
56	842	39	158	51005	43	48995	163	4	837	4
57	881	39	119	048	44	952	167	4	833	3
58	920	39	080	092	43	908	171	4	829	2
59	959	39	041	135	43	865	175	4	825	1
60	48998		51002	51178		48822	02179		97821	0
′	9. l cos	d 1′	10. l sec	9. l cot	d 1′	10. l tan	10. l csc	d 1′	9. l sin	′

Proportional Parts

″	45	44	43	42	41	40	39	5	4
0	0	0	0	0	0	0	0	0	0
1	1	1	1	1	1	1	1	0	0
2	2	1	1	1	1	1	1	0	0
3	2	2	2	2	2	2	2	0	0
4	3	3	3	3	3	3	3	0	0
5	4	4	4	4	3	3	3	0	0
6	4	4	4	4	4	4	4	0	0
7	5	5	5	5	5	5	5	1	0
8	6	6	6	6	5	5	5	1	1
9	7	7	6	6	6	6	6	1	1
10	8	7	7	7	7	7	6	1	1
11	8	8	8	8	8	7	7	1	1
12	9	9	9	8	8	8	8	1	1
13	10	10	9	9	9	9	8	1	1
14	10	10	10	10	10	9	9	1	1
15	11	11	11	10	10	10	10	1	1
16	12	12	11	11	11	11	10	1	1
17	13	12	12	12	12	11	11	1	1
18	14	13	13	13	12	12	12	2	1
19	14	14	14	13	13	13	12	2	1
20	15	15	14	14	14	13	13	2	1
21	16	15	15	15	14	14	14	2	1
22	16	16	16	15	15	15	14	2	1
23	17	17	16	16	16	15	15	2	2
24	18	18	17	17	16	16	16	2	2
25	19	18	18	18	17	17	16	2	2
26	20	19	19	18	18	17	17	2	2
27	20	20	19	19	18	18	18	2	2
28	21	21	20	20	19	19	18	2	2
29	22	21	21	20	20	19	19	2	2
30	22	22	22	21	20	20	20	2	2
31	23	23	22	22	21	21	20	3	2
32	24	23	23	22	22	21	21	3	2
33	25	24	24	23	23	22	21	3	2
34	26	25	24	24	23	23	22	3	2
35	26	26	25	24	24	23	23	3	2
36	27	26	26	25	25	24	23	3	2
37	28	27	27	26	25	25	24	3	2
38	28	28	27	27	26	25	25	3	3
39	29	29	28	27	27	26	25	3	3
40	30	29	29	28	27	27	26	3	3
41	31	30	29	29	28	27	27	3	3
42	32	31	30	29	29	28	27	4	3
43	32	32	31	30	29	29	28	4	3
44	33	32	32	31	30	29	29	4	3
45	34	33	32	32	31	30	29	4	3
46	34	34	33	32	31	31	30	4	3
47	35	34	34	33	32	31	31	4	3
48	36	35	34	34	33	32	31	4	3
49	37	36	35	34	33	33	32	4	3
50	38	37	36	35	34	33	32	4	3
51	38	37	37	36	35	34	33	4	3
52	39	38	37	36	36	35	34	4	3
53	40	39	38	37	36	35	34	4	4
54	40	40	39	38	37	36	35	4	4
55	41	40	39	38	38	37	36	5	4
56	42	41	40	39	38	37	36	5	4
57	43	42	41	40	39	38	37	5	4
58	44	43	42	41	40	39	38	5	4
59	44	43	42	41	40	39	38	5	4
60	45	44	43	42	41	40	39	5	4
″	45	44	43	42	41	40	39	5	4

Proportional Parts

′	l sin 9.	d 1′	l csc 10.	l tan 9.	d 1′	l cot 10.	l sec 10.	d 1′	l cos 9.	′
0	48998	39	51002	51178	43	48822	02179	4	97821	60
1	49037	39	50963	221	43	779	183	5	817	59
2	076	39	924	264	42	736	188	4	812	58
3	115	38	885	306	43	694	192	4	808	57
4	153	39	847	349	43	651	196	4	804	56
5	192	39	808	392	43	608	200	4	800	55
6	231	38	769	435	43	565	204	4	796	54
7	269	39	731	478	42	522	208	4	792	53
8	308	39	692	520	43	480	212	4	788	52
9	347	38	653	563	43	437	216	5	784	51
10	385	39	615	606	42	394	221	4	779	50
11	424	38	576	648	43	352	225	4	775	49
12	462	38	538	691	43	309	229	4	771	48
13	500	39	500	734	42	266	233	4	767	47
14	539	38	461	776	43	224	237	4	763	46
15	577	38	423	819	42	181	241	5	759	45
16	615	39	385	861	42	139	246	4	754	44
17	654	38	346	903	43	097	250	4	750	43
18	692	38	308	946	42	054	254	4	746	42
19	730	38	270	988	43	012	258	4	742	41
20	768	38	232	52031	42	47969	262	4	738	40
21	806	38	194	073	42	927	266	5	734	39
22	844	38	156	115	42	885	271	4	729	38
23	882	38	118	157	43	843	275	4	725	37
24	920	38	080	200	42	800	279	4	721	36
25	958	38	042	242	42	758	283	4	717	35
26	996	38	004	284	42	716	287	5	713	34
27	50034	38	49966	326	42	674	292	4	708	33
28	072	38	928	368	42	632	296	4	704	32
29	110	38	890	410	42	590	300	4	700	31
30	50148	37	49852	52452	42	47548	02304	5	97696	30
31	185	38	815	494	42	506	309	4	691	29
32	223	38	777	536	42	464	313	4	687	28
33	261	37	739	578	42	422	317	4	683	27
34	298	38	702	620	41	380	321	5	679	26
35	336	38	664	661	42	339	326	4	674	25
36	374	37	626	703	42	297	330	4	670	24
37	411	38	589	745	42	255	334	4	666	23
38	449	37	551	787	42	213	338	5	662	22
39	486	37	514	829	41	171	343	4	657	21
40	523	38	477	870	42	130	347	4	653	20
41	561	37	439	912	41	088	351	4	649	19
42	598	37	402	953	42	047	355	5	645	18
43	635	38	365	995	42	005	360	4	640	17
44	673	37	327	53037	41	46963	364	4	636	16
45	710	37	290	078	42	922	368	4	632	15
46	747	37	253	120	41	880	372	5	628	14
47	784	37	216	161	41	839	377	4	623	13
48	821	37	179	202	42	798	381	4	619	12
49	858	38	142	244	41	756	385	5	615	11
50	896	37	104	285	42	715	390	4	610	10
51	933	37	067	327	41	673	394	4	606	9
52	970	37	030	368	41	632	398	5	602	8
53	51007	36	48993	409	41	591	403	4	597	7
54	043	37	957	450	42	550	407	4	593	6
55	080	37	920	492	41	508	411	5	589	5
56	117	37	883	533	41	467	416	4	584	4
57	154	37	846	574	41	426	420	4	580	3
58	191	36	809	615	41	385	424	5	576	2
59	227	37	773	656	41	344	429	4	571	1
60	51264		48736	53697		46303	02433		97567	0
′	9. l cos	d 1′	10. l sec	9. l cot	d 1′	10. l tan	10. l csc	d 1′	9. l sin	′

Proportional Parts

″	43	42	41	39	38	37	36	5	4
0	0	0	0	0	0	0	0	0	0
1	1	1	1	1	1	1	1	0	0
2	1	1	1	1	1	1	1	0	0
3	2	2	2	2	2	2	2	0	0
4	3	3	3	3	3	2	2	0	0
5	4	4	3	3	3	3	3	0	0
6	4	4	4	4	4	4	4	0	0
7	5	5	5	5	4	4	4	1	0
8	6	6	5	5	5	5	5	1	1
9	6	6	6	6	6	6	5	1	1
10	7	7	7	6	6	6	6	1	1
11	8	8	8	7	7	7	7	1	1
12	9	8	8	8	8	7	7	1	1
13	9	9	9	8	8	8	8	1	1
14	10	10	10	9	9	9	8	1	1
15	11	10	10	10	10	9	9	1	1
16	11	11	11	10	10	10	10	1	1
17	12	12	12	11	11	10	10	1	1
18	13	13	12	12	11	11	11	2	1
19	14	13	13	12	12	12	11	2	1
20	14	14	14	13	13	12	12	2	1
21	15	15	14	14	13	13	13	2	1
22	16	15	15	14	14	14	13	2	1
23	16	16	16	15	15	14	14	2	2
24	17	17	16	16	15	15	14	2	2
25	18	18	17	16	16	15	15	2	2
26	19	18	18	17	16	16	16	2	2
27	19	19	18	18	17	17	16	2	2
28	20	20	19	18	18	17	17	2	2
29	21	20	20	19	18	18	17	2	2
30	22	21	20	20	19	18	18	2	2
31	22	22	21	20	20	19	19	3	2
32	23	22	22	21	20	20	19	3	2
33	24	23	23	21	21	20	20	3	2
34	24	24	23	22	22	21	20	3	2
35	25	24	24	23	22	22	21	3	2
36	26	25	25	23	23	22	22	3	2
37	27	26	25	24	23	23	22	3	2
38	27	27	26	25	24	23	23	3	3
39	28	27	27	25	25	24	23	3	3
40	29	28	27	26	25	25	24	3	3
41	29	29	28	27	26	25	25	3	3
42	30	29	29	27	27	26	25	4	3
43	31	30	29	28	27	27	26	4	3
44	32	31	30	29	28	27	26	4	3
45	32	32	31	29	28	28	27	4	3
46	33	32	31	30	29	28	28	4	3
47	34	33	32	31	30	29	28	4	3
48	34	34	33	31	30	30	29	4	3
49	35	34	33	32	31	30	29	4	3
50	36	35	34	32	32	31	30	4	3
51	37	36	35	33	32	31	31	4	3
52	37	36	36	34	33	32	31	4	3
53	38	37	36	34	34	33	32	4	4
54	39	38	37	35	34	33	32	4	4
55	39	38	38	36	35	34	33	5	4
56	40	39	38	36	35	35	34	5	4
57	41	40	39	37	36	35	34	5	4
58	42	41	40	38	37	36	35	5	4
59	42	41	40	38	37	36	35	5	4
60	43	42	41	39	38	37	36	5	4
″	43	42	41	39	38	37	36	5	4

Proportional Parts

′	*l* sin 9.	d 1′	*l* csc 10.	*l* tan 9.	d 1′	*l* cot 10.	*l* sec 10.	d 1′	*l* cos 9.	′
0	51264	37	48736	53697	41	46303	02433	4	97567	60
1	301	37	699	738	41	262	437	5	563	59
2	338	36	662	779	41	221	442	4	558	58
3	374	37	626	820	41	180	446	4	554	57
4	411	36	589	861	41	139	450	5	550	56
5	447	37	553	902	41	098	455	4	545	55
6	484	36	516	943	41	057	459	5	541	54
7	520	37	480	984	41	016	464	4	536	53
8	557	36	443	54025	40	45975	468	4	532	52
9	593	36	407	065	41	935	472	5	528	51
10	629	37	371	106	41	894	477	4	523	50
11	666	36	334	147	40	853	481	4	519	49
12	702	36	298	187	41	813	485	5	515	48
13	738	36	262	228	41	772	490	4	510	47
14	774	37	226	269	40	731	494	5	506	46
15	811	36	189	309	41	691	499	4	501	45
16	847	36	153	350	40	650	503	5	497	44
17	883	36	117	390	41	610	508	4	492	43
18	919	36	081	431	40	569	512	4	488	42
19	955	36	045	471	41	529	516	5	484	41
20	991	36	009	512	40	488	521	4	479	40
21	52027	36	47973	552	41	448	525	5	475	39
22	063	36	937	593	40	407	530	4	470	38
23	099	36	901	633	40	367	534	5	466	37
24	135	36	865	673	41	327	539	4	461	36
25	171	36	829	714	40	286	543	4	457	35
26	207	35	793	754	40	246	547	5	453	34
27	242	36	758	794	41	206	552	4	448	33
28	278	36	722	835	40	165	556	5	444	32
29	314	36	686	875	40	125	561	4	439	31
30	52350	35	47650	54915	40	45085	02565	5	97435	30
31	385	36	615	955	40	045	570	4	430	29
32	421	35	579	995	40	005	574	5	426	28
33	456	36	544	55035	40	44965	579	4	421	27
34	492	35	508	075	40	925	583	5	417	26
35	527	36	473	115	40	885	588	4	412	25
36	563	35	437	155	40	845	592	5	408	24
37	598	36	402	195	40	805	597	4	403	23
38	634	35	366	235	40	765	601	5	399	22
39	669	36	331	275	40	725	606	4	394	21
40	705	35	295	315	40	685	610	5	390	20
41	740	35	260	355	40	645	615	4	385	19
42	775	36	225	395	39	605	619	5	381	18
43	811	35	189	434	40	566	624	4	376	17
44	846	35	154	474	40	526	628	5	372	16
45	881	35	119	514	40	486	633	4	367	15
46	916	35	084	554	39	446	637	5	363	14
47	951	35	049	593	40	407	642	5	358	13
48	986	35	014	633	40	367	647	4	353	12
49	53021	35	46979	673	39	327	651	5	349	11
50	056	36	944	712	40	288	656	4	344	10
51	092	34	908	752	39	248	660	5	340	9
52	126	35	874	791	40	209	665	4	335	8
53	161	35	839	831	39	169	669	5	331	7
54	196	35	804	870	40	130	674	4	326	6
55	231	35	769	910	39	090	678	5	322	5
56	266	35	734	949	40	051	683	5	317	4
57	301	35	699	989	39	011	688	4	312	3
58	336	34	664	56028	39	43972	692	5	308	2
59	370	35	630	067	40	933	697	4	303	1
60	53405		46595	56107		43893	02701		97299	0
′	9. *l* cos	d 1′	10. *l* sec	9. *l* cot	d 1′	10. *l* tan	10. *l* csc	d 1′	9. *l* sin	′

″	Proportional Parts 41	40	39	37	36	35	34	5	4
0	0	0	0	0	0	0	0	0	0
1	1	1	1	1	1	1	1	0	0
2	1	1	1	1	1	1	1	0	0
3	2	2	2	2	2	2	2	0	0
4	3	3	3	2	2	2	2	0	0
5	3	3	3	3	3	3	3	0	0
6	4	4	4	4	4	4	3	0	0
7	5	5	5	4	4	4	4	1	0
8	5	5	5	5	5	5	5	1	1
9	6	6	6	6	5	5	5	1	1
10	7	7	6	6	6	6	6	1	1
11	8	7	7	7	7	6	6	1	1
12	8	8	8	7	7	7	7	1	1
13	9	9	8	8	8	8	7	1	1
14	10	9	9	9	8	8	8	1	1
15	10	10	10	9	9	9	8	1	1
16	11	11	10	10	10	9	9	1	1
17	12	11	11	10	10	10	10	1	1
18	12	12	12	11	11	10	10	2	1
19	13	13	12	12	11	11	11	2	1
20	14	13	13	12	12	12	11	2	1
21	14	14	14	13	13	12	12	2	1
22	15	15	14	14	13	13	12	2	1
23	16	15	15	14	14	13	13	2	2
24	16	16	16	15	14	14	14	2	2
25	17	17	16	15	15	15	14	2	2
26	18	17	17	16	16	15	15	2	2
27	18	18	18	17	16	16	15	2	2
28	19	19	18	17	17	16	16	2	2
29	20	19	19	18	17	17	16	2	2
30	20	20	20	18	18	18	17	2	2
31	21	21	20	19	19	18	18	3	2
32	22	21	21	20	19	19	18	3	2
33	23	22	21	20	20	19	19	3	2
34	23	23	22	21	20	20	19	3	2
35	24	23	23	22	21	20	20	3	2
36	25	24	23	22	22	21	20	3	2
37	25	25	24	23	22	22	21	3	2
38	26	25	25	23	23	22	22	3	3
39	27	26	25	24	23	23	22	3	3
40	27	27	26	25	24	23	23	3	3
41	28	27	27	25	25	24	23	3	3
42	29	28	27	26	25	24	24	4	3
43	29	29	28	27	26	25	24	4	3
44	30	29	29	27	26	26	25	4	3
45	31	30	29	28	27	26	26	4	3
46	31	31	30	28	28	27	26	4	3
47	32	31	31	29	28	27	27	4	3
48	33	32	31	30	29	28	27	4	3
49	33	33	32	30	29	29	28	4	3
50	34	33	32	31	30	29	28	4	3
51	35	34	33	31	31	30	29	4	3
52	36	35	34	32	31	30	29	4	3
53	36	35	34	33	32	31	30	4	4
54	37	36	35	33	32	32	31	4	4
55	38	37	36	34	33	32	31	5	4
56	38	37	36	35	34	33	32	5	4
57	39	38	37	35	34	33	32	5	4
58	40	39	38	36	35	34	33	5	4
59	40	39	38	36	35	34	33	5	4
60	41	40	39	37	36	35	34	5	4
″	41	40	39	37	36	35	34	5	4

Proportional Parts

′	l sin 9.	d 1′	l csc 10.	l tan 9.	d 1′	l cot 10.	l sec 10.	d 1′	l cos 9.	′
0	53405	35	46595	56107	39	43893	02701	5	97299	60
1	440	35	560	146	39	854	706	5	294	59
2	475	34	525	185	39	815	711	4	289	58
3	509	35	491	224	40	776	715	5	285	57
4	544	34	456	264	39	736	720	4	280	56
5	578	35	422	303	39	697	724	5	276	55
6	613	34	387	342	39	658	729	5	271	54
7	647	35	353	381	39	619	734	4	266	53
8	682	34	318	420	39	580	738	5	262	52
9	716	35	284	459	39	541	743	5	257	51
10	751	34	249	498	39	502	748	4	252	50
11	785	34	215	537	39	463	752	5	248	49
12	819	35	181	576	39	424	757	5	243	48
13	854	34	146	615	39	385	762	4	238	47
14	888	34	112	654	39	346	766	5	234	46
15	922	35	078	693	39	307	771	5	229	45
16	957	34	043	732	39	268	776	4	224	44
17	991	34	009	771	39	229	780	5	220	43
18	54025	34	45975	810	39	190	785	5	215	42
19	059	34	941	849	38	151	790	4	210	41
20	093	34	907	887	39	113	794	5	206	40
21	127	34	873	926	39	074	799	5	201	39
22	161	34	839	965	39	035	804	4	196	38
23	195	34	805	57004	38	42996	808	5	192	37
24	229	34	771	042	39	958	813	5	187	36
25	263	34	737	081	39	919	818	4	182	35
26	297	34	703	120	38	880	822	5	178	34
27	331	34	669	158	39	842	827	5	173	33
28	365	34	635	197	38	803	832	5	168	32
29	399	34	601	235	39	765	837	4	163	31
30	54433	33	45567	57274	38	42726	02841	5	97159	30
31	466	34	534	312	39	688	846	5	154	29
32	500	34	500	351	38	649	851	4	149	28
33	534	33	466	389	39	611	855	5	145	27
34	567	34	433	428	38	572	860	5	140	26
35	601	34	399	466	38	534	865	5	135	25
36	635	33	365	504	39	496	870	4	130	24
37	668	34	332	543	38	457	874	5	126	23
38	702	33	298	581	38	419	879	5	121	22
39	735	34	265	619	39	381	884	5	116	21
40	769	33	231	658	38	342	889	4	111	20
41	802	34	198	696	38	304	893	5	107	19
42	836	33	164	734	38	266	898	5	102	18
43	869	34	131	772	38	228	903	5	097	17
44	903	33	097	810	39	190	908	5	092	16
45	936	33	064	849	38	151	913	4	087	15
46	969	34	031	887	38	113	917	5	083	14
47	55003	33	44997	925	38	075	922	5	078	13
48	036	33	964	963	38	037	927	5	073	12
49	069	33	931	58001	38	41999	932	5	068	11
50	102	34	898	039	38	961	937	4	063	10
51	136	33	864	077	38	923	941	5	059	9
52	169	33	831	115	38	885	946	5	054	8
53	202	33	798	153	38	847	951	5	049	7
54	235	33	765	191	38	809	956	5	044	6
55	268	33	732	229	38	771	961	4	039	5
56	301	33	699	267	37	733	965	5	035	4
57	334	33	666	304	38	696	970	5	030	3
58	367	33	633	342	38	658	975	5	025	2
59	400	33	600	380	38	620	980	5	020	1
60	55433		44567	58418		41582	02985		97015	0
′	9. l cos	d 1′	10. l sec	9. l cot	d 1′	10. l tan	10. l csc	d 1′	9. l sin	′

Proportional Parts

″	40	39	38	37	35	34	33	5	4
0	0	0	0	0	0	0	0	0	0
1	1	1	1	1	1	1	1	0	0
2	1	1	1	1	1	1	1	0	0
3	2	2	2	2	2	2	2	0	0
4	3	3	3	2	2	2	2	0	0
5	3	3	3	3	3	3	3	0	0
6	4	4	4	4	4	3	3	0	0
7	5	5	4	4	4	4	4	1	0
8	5	5	5	5	5	5	4	1	1
9	6	6	6	6	5	5	5	1	1
10	7	6	6	6	6	6	6	1	1
11	7	7	7	7	6	6	6	1	1
12	8	8	8	7	7	7	7	1	1
13	9	8	8	8	8	7	7	1	1
14	9	9	9	9	8	8	8	1	1
15	10	10	10	9	9	8	8	1	1
16	11	10	10	10	9	9	9	1	1
17	11	11	11	10	10	10	9	1	1
18	12	12	11	11	10	10	10	2	1
19	13	12	12	12	11	11	10	2	1
20	13	13	13	12	12	11	11	2	1
21	14	14	13	13	12	12	12	2	1
22	15	14	14	14	13	12	12	2	1
23	15	15	15	14	13	13	13	2	2
24	16	16	15	15	14	14	13	2	2
25	17	16	16	15	15	14	14	2	2
26	17	17	16	16	15	15	14	2	2
27	18	18	17	17	16	15	15	2	2
28	19	18	18	17	16	16	15	2	2
29	19	19	18	18	17	16	16	2	2
30	20	20	19	18	18	17	16	2	2
31	21	20	20	19	18	18	17	3	2
32	21	21	20	20	19	18	18	3	2
33	22	21	21	20	19	19	18	3	2
34	23	22	22	21	20	19	19	3	2
35	23	23	22	22	20	20	19	3	2
36	24	23	23	22	21	20	20	3	2
37	25	24	23	23	22	21	20	3	2
38	25	25	24	23	22	22	21	3	3
39	26	25	25	24	23	22	21	3	3
40	27	26	25	25	23	23	22	3	3
41	27	27	26	25	24	23	23	3	3
42	28	27	27	26	24	24	23	4	3
43	29	28	27	27	25	24	24	4	3
44	29	29	28	27	26	25	24	4	3
45	30	29	28	28	26	26	25	4	3
46	31	30	29	28	27	26	25	4	3
47	31	31	30	29	27	27	26	4	3
48	32	31	30	30	28	27	26	4	3
49	33	32	31	30	29	28	27	4	3
50	33	32	32	31	29	28	28	4	3
51	34	33	32	31	30	29	28	4	3
52	35	34	33	32	30	29	29	4	3
53	35	34	34	33	31	30	29	4	4
54	36	35	34	33	32	31	30	4	4
55	37	36	35	34	32	31	30	5	4
56	37	36	35	35	33	32	31	5	4
57	38	37	36	35	33	32	31	5	4
58	39	38	37	36	34	33	32	5	4
59	39	38	37	36	34	33	32	5	4
60	40	39	38	37	35	34	33	5	4
″	40	39	38	37	35	34	33	5	4

Proportional Parts

′	*l* sin 9.	d 1′	*l* csc 10.	*l* tan 9.	d 1′	*l* cot 10.	*l* sec 10.	d 1′	*l* cos 9.	′
0	**55**433	33	**44**567	**58**418	37	**41**582	**02**985	5	**97**015	**60**
1	466	33	534	455	38	545	990	5	010	59
2	499	33	501	493	38	507	995	4	005	58
3	532	32	468	531	38	469	999	5	001	57
4	564	33	436	569	37	431	**03**004	5	**96**996	56
5	597	33	403	606	38	394	009	5	991	**55**
6	630	33	370	644	37	356	014	5	986	54
7	663	32	337	681	38	319	019	5	981	53
8	695	33	305	719	38	281	024	5	976	52
9	728	33	272	757	37	243	029	5	971	51
10	761	32	239	794	38	206	034	4	966	**50**
11	793	33	207	832	37	168	038	5	962	49
12	826	32	174	869	38	131	043	5	957	48
13	858	33	142	907	37	093	048	5	952	47
14	891	32	109	944	37	056	053	5	947	46
15	923	33	077	981	38	019	058	5	942	**45**
16	956	32	044	**59**019	37	**40**981	063	5	937	44
17	988	33	012	056	38	944	068	5	932	43
18	**56**021	32	**43**979	094	37	906	073	5	927	42
19	053	32	947	131	37	869	078	5	922	41
20	085	33	915	168	37	832	083	5	917	**40**
21	118	32	882	205	38	795	088	5	912	39
22	150	32	850	243	37	757	093	4	907	38
23	182	33	818	280	37	720	097	5	903	37
24	215	32	785	317	37	683	102	5	898	36
25	247	32	753	354	37	646	107	5	893	**35**
26	279	32	721	391	38	609	112	5	888	34
27	311	32	689	429	37	571	117	5	883	33
28	343	32	657	466	37	534	122	5	878	32
29	375	33	625	503	37	497	127	5	873	31
30	**56**408	32	**43**592	**59**540	37	**40**460	**03**132	5	**96**868	**30**
31	440	32	560	577	37	423	137	5	863	29
32	472	32	528	614	37	386	142	5	858	28
33	504	32	496	651	37	349	147	5	853	27
34	536	32	464	688	37	312	152	5	848	26
35	568	31	432	725	37	275	157	5	843	**25**
36	599	32	401	762	37	238	162	5	838	24
37	631	32	369	799	36	201	167	5	833	23
38	663	32	337	835	37	165	172	5	828	22
39	695	32	305	872	37	128	177	5	823	21
40	727	32	273	909	37	091	182	5	818	**20**
41	759	31	241	946	37	054	187	5	813	19
42	790	32	210	983	36	017	192	5	808	18
43	822	32	178	**60**019	37	**39**981	197	5	803	17
44	854	32	146	056	37	944	202	5	798	16
45	886	31	114	093	37	907	207	5	793	**15**
46	917	32	083	130	36	870	212	5	788	14
47	949	31	051	166	37	834	217	5	783	13
48	980	32	020	203	37	797	222	6	778	12
49	**57**012	32	**42**988	240	36	760	228	5	772	11
50	044	31	956	276	37	724	233	5	767	**10**
51	075	32	925	313	36	687	238	5	762	9
52	107	31	893	349	37	651	243	5	757	8
53	138	31	862	386	36	614	248	5	752	7
54	169	32	831	422	37	578	253	5	747	6
55	201	31	799	459	36	541	258	5	742	**5**
56	232	32	768	495	37	505	263	5	737	4
57	264	31	736	532	36	468	268	5	732	3
58	295	31	705	568	37	432	273	5	727	2
59	326	32	674	605	36	395	278	5	722	1
60	**57**358		**42**642	**60**641		**39**359	**03**283		**96**717	**0**
′	9. *l* cos	d 1′	10. *l* sec	9. *l* cot	d 1′	10. *l* tan	10. *l* csc	d 1′	9. *l* sin	′

″	38	37	36	33	32	31	6	5
0	0	0	0	0	0	0	0	0
1	1	1	1	1	1	1	0	0
2	1	1	1	1	1	1	0	0
3	2	2	2	2	2	2	0	0
4	3	2	2	2	2	2	0	0
5	3	3	3	3	3	3	0	0
6	4	4	4	3	3	3	1	0
7	4	4	4	4	4	4	1	1
8	5	5	5	4	4	4	1	1
9	6	6	5	5	5	5	1	1
10	6	6	6	6	5	5	1	1
11	7	7	7	6	6	6	1	1
12	8	7	7	7	6	6	1	1
13	8	8	8	7	7	7	1	1
14	9	9	8	8	7	7	1	1
15	10	9	9	8	8	8	2	1
16	10	10	10	9	9	8	2	1
17	11	10	10	9	9	9	2	1
18	11	11	11	10	10	9	2	2
19	12	12	11	10	10	10	2	2
20	13	12	12	11	11	10	2	2
21	13	13	13	12	11	11	2	2
22	14	14	13	12	12	11	2	2
23	15	14	14	13	12	12	2	2
24	15	15	14	13	13	12	2	2
25	16	15	15	14	13	13	2	2
26	16	16	16	14	14	13	3	2
27	17	17	16	15	14	14	3	2
28	18	17	17	15	15	14	3	2
29	18	18	17	16	15	15	3	2
30	19	18	18	16	16	16	3	2
31	20	19	19	17	17	16	3	3
32	20	20	19	18	17	17	3	3
33	21	20	20	18	18	17	3	3
34	22	21	20	19	18	18	3	3
35	22	22	21	19	19	18	4	3
36	23	22	22	20	19	19	4	3
37	23	23	22	20	20	19	4	3
38	24	23	23	21	20	20	4	3
39	25	24	23	21	21	20	4	3
40	25	25	24	22	21	21	4	3
41	26	25	25	23	22	21	4	3
42	27	26	25	23	22	22	4	4
43	27	27	26	24	23	22	4	4
44	28	27	26	24	23	23	4	4
45	28	28	27	25	24	23	4	4
46	29	28	28	25	25	24	5	4
47	30	29	28	26	25	24	5	4
48	30	30	29	26	26	25	5	4
49	31	30	29	27	26	25	5	4
50	32	31	30	28	27	26	5	4
51	32	31	31	28	27	26	5	4
52	33	32	31	29	28	27	5	4
53	34	33	32	29	28	27	5	4
54	34	33	32	30	29	28	5	4
55	35	34	33	30	29	28	6	5
56	35	35	34	31	30	29	6	5
57	36	35	34	31	30	29	6	5
58	37	36	35	32	31	30	6	5
59	37	36	35	32	31	30	6	5
60	38	37	36	33	32	31	6	5
″	38	37	36	33	32	31	6	5

Proportional Parts

′	*l* sin 9.	d 1′	*l* csc 10.	*l* tan 9.	d 1′	*l* cot 10.	*l* sec 10.	d 1′	*l* cos 9.	′
0	57358	31	42642	60641	36	39359	03283	6	96717	60
1	389	31	611	677	37	323	289	5	711	59
2	420	31	580	714	36	286	294	5	706	58
3	451	31	549	750	36	250	299	5	701	57
4	482	32	518	786	37	214	304	5	696	56
5	514	31	486	823	36	177	309	5	691	55
6	545	31	455	859	36	141	314	5	686	54
7	576	31	424	895	36	105	319	5	681	53
8	607	31	393	931	36	069	324	6	676	52
9	638	31	362	967	37	033	330	5	670	51
10	669	31	331	61004	36	38996	335	5	665	50
11	700	31	300	040	36	960	340	5	660	49
12	731	31	269	076	36	924	345	5	655	48
13	762	31	238	112	36	888	350	5	650	47
14	793	31	207	148	36	852	355	5	645	46
15	824	31	176	184	36	816	360	6	640	45
16	855	30	145	220	36	780	366	5	634	44
17	885	31	115	256	36	744	371	5	629	43
18	916	31	084	292	36	708	376	5	624	42
19	947	31	053	328	36	672	381	5	619	41
20	978	30	022	364	36	636	386	6	614	40
21	58008	31	41992	400	36	600	392	5	608	39
22	039	31	961	436	36	564	397	5	603	38
23	070	31	930	472	36	528	402	5	598	37
24	101	30	899	508	36	492	407	5	593	36
25	131	31	869	544	35	456	412	6	588	35
26	162	30	838	579	36	421	418	5	582	34
27	192	31	808	615	36	385	423	5	577	33
28	223	30	777	651	36	349	428	5	572	32
29	253	31	747	687	35	313	433	5	567	31
30	58284	30	41716	61722	36	38278	03438	6	96562	30
31	314	31	686	758	36	242	444	5	556	29
32	345	30	655	794	36	206	449	5	551	28
33	375	31	625	830	35	170	454	5	546	27
34	406	30	594	865	36	135	459	6	541	26
35	436	31	564	901	35	099	465	5	535	25
36	467	30	533	936	36	064	470	5	530	24
37	497	30	503	972	36	028	475	5	525	23
38	527	30	473	62008	35	37992	480	6	520	22
39	557	31	443	043	36	957	486	5	514	21
40	588	30	412	079	35	921	491	5	509	20
41	618	30	382	114	36	886	496	6	504	19
42	648	30	352	150	35	850	502	5	498	18
43	678	31	322	185	36	815	507	5	493	17
44	709	30	291	221	35	779	512	5	488	16
45	739	30	261	256	36	744	517	6	483	15
46	769	30	231	292	35	708	523	5	477	14
47	799	30	201	327	35	673	528	5	472	13
48	829	30	171	362	36	638	533	6	467	12
49	859	30	141	398	35	602	539	5	461	11
50	889	30	111	433	35	567	544	5	456	10
51	919	30	081	468	36	532	549	6	451	9
52	949	30	051	504	35	496	555	5	445	8
53	979	30	021	539	35	461	560	5	440	7
54	59009	30	40991	574	35	426	565	6	435	6
55	039	30	961	609	36	391	571	5	429	5
56	069	29	931	645	35	355	576	5	424	4
57	098	30	902	680	35	320	581	6	419	3
58	128	30	872	715	35	285	587	5	413	2
59	158	30	842	750	35	250	592	5	408	1
60	59188		40812	62785		37215	03597		96403	0
′	9. *l* cos	d 1′	10. *l* sec	9. *l* cot	d 1′	10. *l* tan	10. *l* csc	d 1′	9. *l* sin	′

″	Proportional Parts 37	36	35	32	31	30	29	6	5
0	0	0	0	0	0	0	0	0	0
1	1	1	1	1	1	0	0	0	0
2	1	1	1	1	1	1	1	0	0
3	2	2	2	2	2	2	1	0	0
4	2	2	2	2	2	2	2	0	0
5	3	3	3	3	3	2	2	0	0
6	4	4	4	3	3	3	3	1	0
7	4	4	4	4	4	4	3	1	1
8	5	5	5	4	4	4	4	1	1
9	6	5	5	5	5	4	4	1	1
10	6	6	6	5	5	5	5	1	1
11	7	7	6	6	6	6	5	1	1
12	7	7	7	6	6	6	6	1	1
13	8	8	8	7	7	6	6	1	1
14	9	8	8	7	7	7	7	1	1
15	9	9	9	8	8	8	7	2	1
16	10	10	9	9	8	8	8	2	1
17	10	10	10	9	9	8	8	2	1
18	11	11	10	10	9	9	9	2	2
19	12	11	11	10	10	10	9	2	2
20	12	12	12	11	10	10	10	2	2
21	13	13	12	11	11	10	10	2	2
22	14	13	13	12	11	11	11	2	2
23	14	14	13	12	12	12	11	2	2
24	15	14	14	13	12	12	12	2	2
25	15	15	15	13	13	12	12	2	2
26	16	16	15	14	13	13	13	3	2
27	17	16	16	14	14	14	13	3	2
28	17	17	16	15	14	14	14	3	2
29	18	17	17	15	15	14	14	3	2
30	18	18	18	16	16	15	14	3	2
31	19	19	18	17	16	16	15	3	3
32	20	19	19	17	17	16	15	3	3
33	20	20	19	18	17	16	16	3	3
34	21	20	20	18	18	17	16	3	3
35	22	21	20	19	18	18	17	4	3
36	22	22	21	19	19	18	17	4	3
37	23	22	22	20	19	18	18	4	3
38	23	23	22	20	20	19	18	4	3
39	24	23	23	21	20	20	19	4	3
40	25	24	23	21	21	20	19	4	3
41	25	25	24	22	21	20	20	4	3
42	26	25	24	22	22	21	20	4	4
43	27	26	25	23	22	22	21	4	4
44	27	26	26	23	23	22	21	4	4
45	28	27	26	24	23	22	22	4	4
46	28	28	27	25	24	23	22	5	4
47	29	28	27	25	24	24	23	5	4
48	30	29	28	26	25	24	23	5	4
49	30	29	29	26	25	24	24	5	4
50	31	30	29	27	26	25	24	5	4
51	31	31	30	27	26	26	25	5	4
52	32	31	30	28	27	26	25	5	4
53	33	32	31	28	27	26	26	5	4
54	33	32	32	29	28	27	26	5	4
55	34	33	32	29	28	28	27	6	5
56	35	34	33	30	29	28	27	6	5
57	35	34	33	30	29	28	28	6	5
58	36	35	34	31	30	29	28	6	5
59	36	35	34	31	30	30	29	6	5
60	37	36	35	32	31	30	29	6	5
″	37	36	35	32	31	30	29	6	5

Proportional Parts

′	*l* sin 9.	d 1′	*l* csc 10.	*l* tan 9.	d 1′	*l* cot 10.	*l* sec 10.	d 1′	*l* cos 9.	′
0	59188	30	40812	62785	35	37215	03597	6	96403	60
1	218	29	782	820	35	180	603	5	397	59
2	247	30	753	855	35	145	608	5	392	58
3	277	30	723	890	36	110	613	6	387	57
4	307	29	693	926	35	074	619	5	381	56
5	336	30	664	961	35	039	624	6	376	55
6	366	30	634	996	35	004	630	5	370	54
7	396	29	604	63031	35	36969	635	5	365	53
8	425	30	575	066	35	934	640	6	360	52
9	455	29	545	101	34	899	646	5	354	51
10	484	30	516	135	35	865	651	6	349	50
11	514	29	486	170	35	830	657	5	343	49
12	543	30	457	205	35	795	662	5	338	48
13	573	29	427	240	35	760	667	6	333	47
14	602	30	398	275	35	725	673	5	327	46
15	632	29	368	310	35	690	678	6	322	45
16	661	29	339	345	34	655	684	5	316	44
17	690	30	310	379	35	621	689	6	311	43
18	720	29	280	414	35	586	695	5	305	42
19	749	29	251	449	35	551	700	6	300	41
20	778	30	222	484	35	516	706	5	294	40
21	808	29	192	519	34	481	711	5	289	39
22	837	29	163	553	35	447	716	6	284	38
23	866	29	134	588	35	412	722	5	278	37
24	895	29	105	623	34	377	727	6	273	36
25	924	30	076	657	35	343	733	5	267	35
26	954	29	046	692	34	308	738	6	262	34
27	983	29	017	726	35	274	744	5	256	33
28	60012	29	39988	761	35	239	749	6	251	32
29	041	29	959	796	34	204	755	5	245	31
30	60070	29	39930	63830	35	36170	03760	6	96240	30
31	099	29	901	865	34	135	766	5	234	29
32	128	29	872	899	35	101	771	6	229	28
33	157	29	843	934	34	066	777	5	223	27
34	186	29	814	968	35	032	782	6	218	26
35	215	29	785	64003	34	35997	788	5	212	25
36	244	29	756	037	35	963	793	6	207	24
37	273	29	727	072	34	928	799	5	201	23
38	302	29	698	106	34	894	804	6	196	22
39	331	28	669	140	35	860	810	5	190	21
40	359	29	641	175	34	825	815	6	185	20
41	388	29	612	209	34	791	821	5	179	19
42	417	29	583	243	35	757	826	6	174	18
43	446	28	554	278	34	722	832	6	168	17
44	474	29	526	312	34	688	838	5	162	16
45	503	29	497	346	35	654	843	6	157	15
46	532	29	468	381	34	619	849	5	151	14
47	561	28	439	415	34	585	854	6	146	13
48	589	29	411	449	34	551	860	5	140	12
49	618	28	382	483	34	517	865	6	135	11
50	646	29	354	517	35	483	871	6	129	10
51	675	29	325	552	34	448	877	5	123	9
52	704	28	296	586	34	414	882	6	118	8
53	732	29	268	620	34	380	888	5	112	7
54	761	28	239	654	34	346	893	6	107	6
55	789	29	211	688	34	312	899	6	101	5
56	818	28	182	722	34	278	905	5	095	4
57	846	29	154	756	34	244	910	6	090	3
58	875	28	125	790	34	210	916	5	084	2
59	903	28	097	824	34	176	921	6	079	1
60	60931		39069	64858		35142	03927		96073	0
′	9. *l* cos	d 1′	10. *l* sec	9. *l* cot	d 1′	10. *l* tan	10. *l* csc	d 1′	9. *l* sin	′

113° 66°

″	Proportional Parts 36	35	34	30	29	28	6	5
0	0	0	0	0	0	0	0	0
1	1	1	1	0	0	0	0	0
2	1	1	1	1	1	1	0	0
3	2	2	2	2	1	1	0	0
4	2	2	2	2	2	2	0	0
5	3	3	3	2	2	2	0	0
6	4	4	3	3	3	3	1	0
7	4	4	4	4	3	3	1	1
8	5	5	5	4	4	4	1	1
9	5	5	5	4	4	4	1	1
10	6	6	6	5	5	5	1	1
11	7	6	6	6	5	5	1	1
12	7	7	7	6	6	6	1	1
13	8	8	7	6	6	6	1	1
14	8	8	8	7	7	7	1	1
15	9	9	8	8	7	7	2	1
16	10	9	9	8	8	7	2	1
17	10	10	10	8	8	8	2	1
18	11	10	10	9	9	8	2	2
19	11	11	11	10	9	9	2	2
20	12	12	11	10	10	9	2	2
21	13	12	12	10	10	10	2	2
22	13	13	12	11	11	10	2	2
23	14	13	13	12	11	11	2	2
24	14	14	14	12	12	11	2	2
25	15	15	14	12	12	12	2	2
26	16	15	15	13	13	12	3	2
27	16	16	15	14	13	13	3	2
28	17	16	16	14	14	13	3	2
29	17	17	16	14	14	14	3	2
30	18	18	17	15	14	14	3	2
31	19	18	18	16	15	14	3	3
32	19	19	18	16	15	15	3	3
33	20	19	19	16	16	15	3	3
34	20	20	19	17	16	16	3	3
35	21	20	20	18	17	16	4	3
36	22	21	20	18	17	17	4	3
37	22	22	21	18	18	17	4	3
38	23	22	22	19	18	18	4	3
39	23	23	22	20	19	18	4	3
40	24	23	23	20	19	19	4	3
41	25	24	23	20	20	19	4	3
42	25	24	24	21	20	20	4	4
43	26	25	24	22	21	20	4	4
44	26	26	25	22	21	21	4	4
45	27	26	26	22	22	21	4	4
46	28	27	26	23	22	21	5	4
47	28	27	27	24	23	22	5	4
48	29	28	27	24	23	22	5	4
49	29	29	28	24	24	23	5	4
50	30	29	28	25	24	23	5	4
51	31	30	29	26	25	24	5	4
52	31	30	29	26	25	24	5	4
53	32	31	30	26	26	25	5	4
54	32	32	31	27	26	25	5	4
55	33	32	31	28	27	26	6	5
56	34	33	32	28	27	26	6	5
57	34	33	32	28	28	27	6	5
58	35	34	33	29	28	27	6	5
59	35	34	33	30	29	28	6	5
60	36	35	34	30	29	28	6	5
″	36 Proportional Parts	35	34	30	29	28	6	5

′	l sin 9.	d 1′	l csc 10.	l tan 9.	d 1′	l cot 10.	l sec 10.	d 1′	l cos 9.	′
0	60931	29	39069	64858	34	35142	03927	6	96073	60
1	960	28	040	892	34	108	933	5	067	59
2	988	28	012	926	34	074	938	6	062	58
3	61016	29	38984	960	34	040	944	6	056	57
4	045	28	955	994	34	006	950	5	050	56
5	073	28	927	65028	34	34972	955	6	045	55
6	101	28	899	062	34	938	961	5	039	54
7	129	29	871	096	34	904	966	6	034	53
8	158	28	842	130	34	870	972	6	028	52
9	186	28	814	164	33	836	978	5	022	51
10	214	28	786	197	34	803	983	6	017	50
11	242	28	758	231	34	769	989	6	011	49
12	270	28	730	265	34	735	995	5	005	48
13	298	28	702	299	34	701	04000	6	000	47
14	326	28	674	333	33	667	006	6	95994	46
15	354	28	646	366	34	634	012	6	988	45
16	382	29	618	400	34	600	018	5	982	44
17	411	27	589	434	33	566	023	6	977	43
18	438	28	562	467	34	533	029	6	971	42
19	466	28	534	501	34	499	035	5	965	41
20	494	28	506	535	33	465	040	6	960	40
21	522	28	478	568	34	432	046	6	954	39
22	550	28	450	602	34	398	052	6	948	38
23	578	28	422	636	33	364	058	5	942	37
24	606	28	394	669	34	331	063	6	937	36
25	634	28	366	703	33	297	069	6	931	35
26	662	27	338	736	34	264	075	5	925	34
27	689	28	311	770	33	230	080	6	920	33
28	717	28	283	803	34	197	086	6	914	32
29	745	28	255	837	33	163	092	6	908	31
30	61773	27	38227	65870	34	34130	04098	5	95902	30
31	800	28	200	904	33	096	103	6	897	29
32	828	28	172	937	34	063	109	6	891	28
33	856	27	144	971	33	029	115	6	885	27
34	883	28	117	66004	34	33996	121	6	879	26
35	911	28	089	038	33	962	127	5	873	25
36	939	27	061	071	33	929	132	6	868	24
37	966	28	034	104	34	896	138	6	862	23
38	994	27	006	138	33	862	144	6	856	22
39	62021	28	37979	171	33	829	150	6	850	21
40	049	27	951	204	34	796	156	5	844	20
41	076	28	924	238	33	762	161	6	839	19
42	104	27	896	271	33	729	167	6	833	18
43	131	28	869	304	33	696	173	6	827	17
44	159	27	841	337	34	663	179	6	821	16
45	186	28	814	371	33	629	185	5	815	15
46	214	27	786	404	33	596	190	6	810	14
47	241	27	759	437	33	563	196	6	804	13
48	268	28	732	470	33	530	202	6	798	12
49	296	27	704	503	34	497	208	6	792	11
50	323	27	677	537	33	463	214	6	786	10
51	350	27	650	570	33	430	220	5	780	9
52	377	28	623	603	33	397	225	6	775	8
53	405	27	595	636	33	364	231	6	769	7
54	432	27	568	669	33	331	237	6	763	6
55	459	27	541	702	33	298	243	6	757	5
56	486	27	514	735	33	265	249	6	751	4
57	513	28	487	768	33	232	255	6	745	3
58	541	27	459	801	33	199	261	6	739	2
59	568	27	432	834	33	166	267	5	733	1
60	62595		37405	66867		33133	04272		95728	0
′	9. l cos	d 1′	10. l sec	9. l cot	d 1′	10. l tan	10. l csc	d 1′	9. l sin	′

″	Proportional Parts 34	33	29	28	27	6	5
0	0	0	0	0	0	0	0
1	1	1	0	0	0	0	0
2	1	1	1	1	1	0	0
3	2	2	1	1	1	0	0
4	2	2	2	2	2	0	0
5	3	3	2	2	2	0	0
6	3	3	3	3	3	1	0
7	4	4	3	3	3	1	1
8	5	4	4	4	4	1	1
9	5	5	4	4	4	1	1
10	6	6	5	5	4	1	1
11	6	6	5	5	5	1	1
12	7	7	6	6	5	1	1
13	7	7	6	6	6	1	1
14	8	8	7	7	6	1	1
15	8	8	7	7	7	2	1
16	9	9	8	7	7	2	1
17	10	9	8	8	8	2	1
18	10	10	9	8	8	2	2
19	11	10	9	9	9	2	2
20	11	11	10	9	9	2	2
21	12	12	10	10	9	2	2
22	12	12	11	10	10	2	2
23	13	13	11	11	10	2	2
24	14	13	12	11	11	2	2
25	14	14	12	12	11	2	2
26	15	14	13	12	12	3	2
27	15	15	13	13	12	3	2
28	16	15	14	13	13	3	2
29	16	16	14	14	13	3	2
30	17	16	14	14	14	3	2
31	18	17	15	14	14	3	3
32	18	18	15	15	14	3	3
33	19	18	16	15	15	3	3
34	19	19	16	16	15	3	3
35	20	19	17	16	16	4	3
36	20	20	17	17	16	4	3
37	21	20	18	17	17	4	3
38	22	21	18	18	17	4	3
39	22	21	19	18	18	4	3
40	23	22	19	19	18	4	3
41	23	23	20	19	18	4	3
42	24	23	20	20	19	4	4
43	24	24	21	20	19	4	4
44	25	24	21	21	20	4	4
45	26	25	22	21	20	4	4
46	26	25	22	21	21	5	4
47	27	26	23	22	21	5	4
48	27	26	23	22	22	5	4
49	28	27	24	23	22	5	4
50	28	28	24	23	22	5	4
51	29	28	25	24	23	5	4
52	29	29	25	24	23	5	4
53	30	29	26	25	24	5	4
54	31	30	26	25	24	5	4
55	31	30	27	26	25	6	5
56	32	31	27	26	25	6	5
57	32	31	28	27	26	6	5
58	33	32	28	27	26	6	5
59	33	32	29	28	27	6	5
60	34	33	29	28	27	6	5
″	34	33	29	28	27	6	5

Proportional Parts

′	l sin 9.	d 1′	l csc 10.	l tan 9.	d 1′	l cot 10.	l sec 10.	d 1′	l cos 9.	′
0	62595	27	37405	66867	33	33133	04272	6	95728	60
1	622	27	378	900	33	100	278	6	722	59
2	649	27	351	933	33	067	284	6	716	58
3	676	27	324	966	33	034	290	6	710	57
4	703	27	297	999	33	001	296	6	704	56
5	730	27	270	67032	33	32968	302	6	698	55
6	757	27	243	065	33	935	308	6	692	54
7	784	27	216	098	33	902	314	6	686	53
8	811	27	189	131	32	869	320	6	680	52
9	838	27	162	163	33	837	326	6	674	51
10	865	27	135	196	33	804	332	5	668	50
11	892	26	108	229	33	771	337	6	663	49
12	918	27	082	262	33	738	343	6	657	48
13	945	27	055	295	32	705	349	6	651	47
14	972	27	028	327	33	673	355	6	645	46
15	999	27	001	360	33	640	361	6	639	45
16	63026	26	36974	393	33	607	367	6	633	44
17	052	27	948	426	32	574	373	6	627	43
18	079	27	921	458	33	542	379	6	621	42
19	106	27	894	491	33	509	385	6	615	41
20	133	26	867	524	32	476	391	6	609	40
21	159	27	841	556	33	444	397	6	603	39
22	186	27	814	589	33	411	403	6	597	38
23	213	26	787	622	32	378	409	6	591	37
24	239	27	761	654	33	346	415	6	585	36
25	266	26	734	687	32	313	421	6	579	35
26	292	27	708	719	33	281	427	6	573	34
27	319	26	681	752	33	248	433	6	567	33
28	345	27	655	785	32	215	439	6	561	32
29	372	26	628	817	33	183	445	6	555	31
30	63398	27	36602	67850	32	32150	04451	6	95549	30
31	425	26	575	882	33	118	457	6	543	29
32	451	27	549	915	32	085	463	6	537	28
33	478	26	522	947	33	053	469	6	531	27
34	504	27	496	980	32	020	475	6	525	26
35	531	26	469	68012	32	31988	481	6	519	25
36	557	26	443	044	33	956	487	6	513	24
37	583	27	417	077	32	923	493	7	507	23
38	610	26	390	109	33	891	500	6	500	22
39	636	26	364	142	32	858	506	6	494	21
40	662	27	338	174	32	826	512	6	488	20
41	689	26	311	206	33	794	518	6	482	19
42	715	26	285	239	32	761	524	6	476	18
43	741	26	259	271	32	729	530	6	470	17
44	767	27	233	303	33	697	536	6	464	16
45	794	26	206	336	32	664	542	6	458	15
46	820	26	180	368	32	632	548	6	452	14
47	846	26	154	400	32	600	554	6	446	13
48	872	26	128	432	33	568	560	6	440	12
49	898	26	102	465	32	535	566	7	434	11
50	924	26	076	497	32	503	573	6	427	10
51	950	26	050	529	32	471	579	6	421	9
52	976	26	024	561	32	439	585	6	415	8
53	64002	26	35998	593	33	407	591	6	409	7
54	028	26	972	626	32	374	597	6	403	6
55	054	26	946	658	32	342	603	6	397	5
56	080	26	920	690	32	310	609	7	391	4
57	106	26	894	722	32	278	616	6	384	3
58	132	26	868	754	32	246	622	6	378	2
59	158	26	842	786	32	214	628	6	372	1
60	64184		35816	68818		31182	04634		95366	0
′	9. l cos	d 1′	10. l sec	9. l cot	d 1′	10. l tan	10. l csc	d 1′	9. l sin	′

″	Proportional Parts 33	32	27	26	7	6	5
0	0	0	0	0	0	0	0
1	1	1	0	0	0	0	0
2	1	1	1	1	0	0	0
3	2	2	1	1	0	0	0
4	2	2	2	2	0	0	0
5	3	3	2	2	1	0	0
6	3	3	3	3	1	1	0
7	4	4	3	3	1	1	1
8	4	4	4	3	1	1	1
9	5	5	4	4	1	1	1
10	6	5	4	4	1	1	1
11	6	6	5	5	1	1	1
12	7	6	5	5	1	1	1
13	7	7	6	6	2	1	1
14	8	7	6	6	2	1	1
15	8	8	7	6	2	2	1
16	9	9	7	7	2	2	1
17	9	9	8	7	2	2	1
18	10	10	8	8	2	2	2
19	10	10	9	8	2	2	2
20	11	11	9	9	2	2	2
21	12	11	9	9	2	2	2
22	12	12	10	10	3	2	2
23	13	12	10	10	3	2	2
24	13	13	11	10	3	2	2
25	14	13	11	11	3	2	2
26	14	14	12	11	3	3	2
27	15	14	12	12	3	3	2
28	15	15	13	12	3	3	2
29	16	15	13	13	3	3	2
30	16	16	14	13	4	3	2
31	17	17	14	13	4	3	3
32	18	17	14	14	4	3	3
33	18	18	15	14	4	3	3
34	19	18	15	15	4	3	3
35	19	19	16	15	4	4	3
36	20	19	16	16	4	4	3
37	20	20	17	16	4	4	3
38	21	20	17	16	4	4	3
39	21	21	18	17	5	4	3
40	22	21	18	17	5	4	3
41	23	22	18	18	5	4	3
42	23	22	19	18	5	4	4
43	24	23	19	19	5	4	4
44	24	23	20	19	5	4	4
45	25	24	20	20	5	4	4
46	25	25	21	20	5	5	4
47	26	25	21	20	5	5	4
48	26	26	22	21	6	5	4
49	27	26	22	21	6	5	4
50	28	27	22	22	6	5	4
51	28	27	23	22	6	5	4
52	29	28	23	23	6	5	4
53	29	28	24	23	6	5	4
54	30	29	24	23	6	5	4
55	30	29	25	24	6	6	5
56	31	30	25	24	7	6	5
57	31	30	26	25	7	6	5
58	32	31	26	25	7	6	5
59	32	31	27	26	7	6	5
60	33	32	27	26	7	6	5
″	33 Proportional Parts	32	27	26	7	6	5

′	l sin 9.	d 1′	l csc 10.	l tan 9.	d 1′	l cot 10.	l sec 10.	d 1′	l cos 9.	′
0	64184	26	35816	68818	32	31182	04634	6	95366	60
1	210	26	790	850	32	150	640	6	360	59
2	236	26	764	882	32	118	646	6	354	58
3	262	26	738	914	32	086	652	7	348	57
4	288	25	712	946	32	054	659	6	341	56
5	313	26	687	978	32	022	665	6	335	55
6	339	26	661	69010	32	30990	671	6	329	54
7	365	26	635	042	32	958	677	6	323	53
8	391	26	609	074	32	926	683	7	317	52
9	417	25	583	106	32	894	690	6	310	51
10	442	26	558	138	32	862	696	6	304	50
11	468	26	532	170	32	830	702	6	298	49
12	494	25	506	202	32	798	708	6	292	48
13	519	26	481	234	32	766	714	7	286	47
14	545	26	455	266	32	734	721	6	279	46
15	571	25	429	298	31	702	727	6	273	45
16	596	26	404	329	32	671	733	6	267	44
17	622	25	378	361	32	639	739	7	261	43
18	647	26	353	393	32	607	746	6	254	42
19	673	25	327	425	32	575	752	6	248	41
20	698	26	302	457	31	543	758	6	242	40
21	724	25	276	488	32	512	764	7	236	39
22	749	26	251	520	32	480	771	6	229	38
23	775	25	225	552	32	448	777	6	223	37
24	800	26	200	584	31	416	783	6	217	36
25	826	25	174	615	32	385	789	7	211	35
26	851	26	149	647	32	353	796	6	204	34
27	877	25	123	679	31	321	802	6	198	33
28	902	25	098	710	32	290	808	7	192	32
29	927	26	073	742	32	258	815	6	185	31
30	64953	25	35047	69774	31	30226	04821	6	95179	30
31	978	25	022	805	32	195	827	6	173	29
32	65003	26	34997	837	31	163	833	7	167	28
33	029	25	971	868	32	132	840	6	160	27
34	054	25	946	900	32	100	846	6	154	26
35	079	25	921	932	31	068	852	7	148	25
36	104	26	896	963	32	037	859	6	141	24
37	130	25	870	995	31	005	865	6	135	23
38	155	25	845	70026	32	29974	871	7	129	22
39	180	25	820	058	31	942	878	6	122	21
40	205	25	795	089	32	911	884	6	116	20
41	230	25	770	121	31	879	890	7	110	19
42	255	26	745	152	32	848	897	6	103	18
43	281	25	719	184	31	816	903	7	097	17
44	306	25	694	215	32	785	910	6	090	16
45	331	25	669	247	31	753	916	6	084	15
46	356	25	644	278	31	722	922	7	078	14
47	381	25	619	309	32	691	929	6	071	13
48	406	25	594	341	31	659	935	6	065	12
49	431	25	569	372	32	628	941	7	059	11
50	456	25	544	404	31	596	948	6	052	10
51	481	25	519	435	31	565	954	7	046	9
52	506	25	494	466	32	534	961	6	039	8
53	531	25	469	498	31	502	967	6	033	7
54	556	24	444	529	31	471	973	7	027	6
55	580	25	420	560	32	440	980	6	020	5
56	605	25	395	592	31	408	986	7	014	4
57	630	25	370	623	31	377	993	6	007	3
58	655	25	345	654	31	346	999	6	001	2
59	680	25	320	685	32	315	05005	7	94995	1
60	65705		34295	70717		29283	05012		94988	0
′	9. l cos	d 1′	10. l sec	9. l cot	d 1′	10. l tan	10. l csc	d 1′	9. l sin	′

″	Proportional Parts 32	31	26	25	24	7	6
0	0	0	0	0	0	0	0
1	1	1	0	0	0	0	0
2	1	1	1	1	1	0	0
3	2	2	1	1	1	0	0
4	2	2	2	2	2	0	0
5	3	3	2	2	2	1	0
6	3	3	3	2	2	1	1
7	4	4	3	3	3	1	1
8	4	4	3	3	3	1	1
9	5	5	4	4	4	1	1
10	5	5	4	4	4	1	1
11	6	6	5	5	4	1	1
12	6	6	5	5	5	1	1
13	7	7	6	5	5	2	1
14	7	7	6	6	6	2	1
15	8	8	6	6	6	2	2
16	9	8	7	7	6	2	2
17	9	9	7	7	7	2	2
18	10	9	8	8	7	2	2
19	10	10	8	8	8	2	2
20	11	10	9	8	8	2	2
21	11	11	9	9	8	2	2
22	12	11	10	9	9	3	2
23	12	12	10	10	9	3	2
24	13	12	10	10	10	3	2
25	13	13	11	10	10	3	2
26	14	13	11	11	10	3	3
27	14	14	12	11	11	3	3
28	15	14	12	12	11	3	3
29	15	15	13	12	12	3	3
30	16	16	13	12	12	4	3
31	17	16	13	13	12	4	3
32	17	17	14	13	13	4	3
33	18	17	14	14	13	4	3
34	18	18	15	14	14	4	3
35	19	18	15	15	14	4	4
36	19	19	16	15	14	4	4
37	20	19	16	15	15	4	4
38	20	20	16	16	15	4	4
39	21	20	17	16	16	5	4
40	21	21	17	17	16	5	4
41	22	21	18	17	16	5	4
42	22	22	18	18	17	5	4
43	23	22	19	18	17	5	4
44	23	23	19	18	18	5	4
45	24	23	20	19	18	5	4
46	25	24	20	19	18	5	5
47	25	24	20	20	19	5	5
48	26	25	21	20	19	6	5
49	26	25	21	20	20	6	5
50	27	26	22	21	20	6	5
51	27	26	22	21	20	6	5
52	28	27	23	22	21	6	5
53	28	27	23	22	21	6	5
54	29	28	23	22	22	6	5
55	29	28	24	23	22	6	6
56	30	29	24	23	22	7	6
57	30	29	25	24	23	7	6
58	31	30	25	24	23	7	6
59	31	30	26	25	24	7	6
60	32	31	26	25	24	7	6
″	32	31	26	25	24	7	6
	Proportional Parts						

′	*l* sin 9.	d 1′	*l* csc 10.	*l* tan 9.	d 1′	*l* cot 10.	*l* sec 10.	d 1′	*l* cos 9.	′
0	65705	24	34295	70717	31	29283	05012	6	94988	60
1	729	25	271	748	31	252	018	7	982	59
2	754	25	246	779	31	221	025	6	975	58
3	779	25	221	810	31	190	031	7	969	57
4	804	24	196	841	32	159	038	6	962	56
5	828	25	172	873	31	127	044	7	956	55
6	853	25	147	904	31	096	051	6	949	54
7	878	24	122	935	31	065	057	7	943	53
8	902	25	098	966	31	034	064	6	936	52
9	927	25	073	997	31	003	070	7	930	51
10	952	24	048	71028	31	28972	077	6	923	50
11	976	25	024	059	31	941	083	6	917	49
12	66001	24	33999	090	31	910	089	7	911	48
13	025	25	975	121	32	879	096	6	904	47
14	050	25	950	153	31	847	102	7	898	46
15	075	24	925	184	31	816	109	6	891	45
16	099	25	901	215	31	785	115	7	885	44
17	124	24	876	246	31	754	122	7	878	43
18	148	25	852	277	31	723	129	6	871	42
19	173	24	827	308	31	692	135	7	865	41
20	197	24	803	339	31	661	142	6	858	40
21	221	25	779	370	31	630	148	7	852	39
22	246	24	754	401	30	599	155	6	845	38
23	270	25	730	431	31	569	161	7	839	37
24	295	24	705	462	31	538	168	6	832	36
25	319	24	681	493	31	507	174	7	826	35
26	343	25	657	524	31	476	181	6	819	34
27	368	24	632	555	31	445	187	7	813	33
28	392	24	608	586	31	414	194	7	806	32
29	416	25	584	617	31	383	201	6	799	31
30	66441	24	33559	71648	31	28352	05207	7	94793	30
31	465	24	535	679	30	321	214	6	786	29
32	489	24	511	709	31	291	220	7	780	28
33	513	24	487	740	31	260	227	6	773	27
34	537	25	463	771	31	229	233	7	767	26
35	562	24	438	802	31	198	240	7	760	25
36	586	24	414	833	30	167	247	6	753	24
37	610	24	390	863	31	137	253	7	747	23
38	634	24	366	894	31	106	260	6	740	22
39	658	24	342	925	30	075	266	7	734	21
40	682	24	318	955	31	045	273	7	727	20
41	706	25	294	986	31	014	280	6	720	19
42	731	24	269	72017	31	27983	286	7	714	18
43	755	24	245	048	30	952	293	7	707	17
44	779	24	221	078	31	922	300	6	700	16
45	803	24	197	109	31	891	306	7	694	15
46	827	24	173	140	30	860	313	7	687	14
47	851	24	149	170	31	830	320	6	680	13
48	875	24	125	201	30	799	326	7	674	12
49	899	23	101	231	31	769	333	7	667	11
50	922	24	078	262	31	738	340	6	660	10
51	946	24	054	293	30	707	346	7	654	9
52	970	24	030	323	31	677	353	7	647	8
53	994	24	006	354	30	646	360	6	640	7
54	67018	24	32982	384	31	616	366	7	634	6
55	042	24	958	415	30	585	373	7	627	5
56	066	24	934	445	31	555	380	6	620	4
57	090	23	910	476	30	524	386	7	614	3
58	113	24	887	506	31	494	393	7	607	2
59	137	24	863	537	30	463	400	7	600	1
60	67161		32839	72567		27433	05407		94593	0
′	9 *l* cos	d 1′	10. *l* sec	9. *l* cot	d 1′	10. *l* tan	10. *l* csc	d 1′	9. *l* sin	′

″	Proportional Parts 32	31	30	25	24	23	7	6
0	0	0	0	0	0	0	0	0
1	1	1	0	0	0	0	0	0
2	1	1	1	1	1	1	0	0
3	2	2	2	1	1	1	0	0
4	2	2	2	2	2	2	0	0
5	3	3	2	2	2	2	1	0
6	3	3	3	2	2	2	1	1
7	4	4	4	3	3	3	1	1
8	4	4	4	3	3	3	1	1
9	5	5	4	4	4	3	1	1
10	5	5	5	4	4	4	1	1
11	6	6	6	5	4	4	1	1
12	6	6	6	5	5	5	1	1
13	7	7	6	5	5	5	2	1
14	7	7	7	6	6	5	2	1
15	8	8	8	6	6	6	2	2
16	9	8	8	7	6	6	2	2
17	9	9	8	7	7	7	2	2
18	10	9	9	8	7	7	2	2
19	10	10	10	8	8	7	2	2
20	11	10	10	8	8	8	2	2
21	11	11	10	9	8	8	2	2
22	12	11	11	9	9	8	3	2
23	12	12	12	10	9	9	3	2
24	13	12	12	10	10	9	3	2
25	13	13	12	10	10	10	3	2
26	14	13	13	11	10	10	3	3
27	14	14	14	11	11	10	3	3
28	15	14	14	12	11	11	3	3
29	15	15	14	12	12	11	3	3
30	16	16	15	12	12	12	4	3
31	17	16	16	13	12	12	4	3
32	17	17	16	13	13	12	4	3
33	18	17	16	14	13	13	4	3
34	18	18	17	14	14	13	4	3
35	19	18	18	15	14	13	4	4
36	19	19	18	15	14	14	4	4
37	20	19	18	15	15	14	4	4
38	20	20	19	16	15	15	4	4
39	21	20	20	16	16	15	5	4
40	21	21	20	17	16	15	5	4
41	22	21	20	17	16	16	5	4
42	22	22	21	18	17	16	5	4
43	23	22	22	18	17	16	5	4
44	23	23	22	18	18	17	5	4
45	24	23	22	19	18	17	5	4
46	25	24	23	19	18	18	5	5
47	25	24	24	20	19	18	5	5
48	26	25	24	20	19	18	6	5
49	26	25	24	20	20	19	6	5
50	27	26	25	21	20	19	6	5
51	27	26	26	21	20	20	6	5
52	28	27	26	22	21	20	6	5
53	28	27	26	22	21	20	6	5
54	29	28	27	22	22	21	6	5
55	29	28	28	23	22	21	6	6
56	30	29	28	23	22	21	7	6
57	30	29	28	24	23	22	7	6
58	31	30	29	24	23	22	7	6
59	31	30	30	25	24	23	7	6
60	32	31	30	25	24	23	7	6
″	32 Proportional Parts	31	30	25	24	23	7	6

′	l sin 9.	d 1′	l csc 10.	l tan 9.	d 1′	l cot 10.	l sec 10.	d 1′	l cos 9.	′
0	67161	24	32839	72567	31	27433	05407	6	94593	60
1	185	23	815	598	30	402	413	7	587	59
2	208	24	792	628	31	372	420	7	580	58
3	232	24	768	659	30	341	427	6	573	57
4	256	24	744	689	31	311	433	7	567	56
5	280	23	720	720	30	280	440	7	560	55
6	303	24	697	750	30	250	447	7	553	54
7	327	23	673	780	31	220	454	6	546	53
8	350	24	650	811	30	189	460	7	540	52
9	374	24	626	841	31	159	467	7	533	51
10	398	23	602	872	30	128	474	7	526	50
11	421	24	579	902	30	098	481	6	519	49
12	445	23	555	932	31	068	487	7	513	48
13	468	24	532	963	30	037	494	7	506	47
14	492	23	508	993	30	007	501	7	499	46
15	515	24	485	73023	31	26977	508	7	492	45
16	539	23	461	054	30	946	515	6	485	44
17	562	24	438	084	30	916	521	7	479	43
18	586	23	414	114	30	886	528	7	472	42
19	609	24	391	144	31	856	535	7	465	41
20	633	23	367	175	30	825	542	7	458	40
21	656	24	344	205	30	795	549	6	451	39
22	680	23	320	235	30	765	555	7	445	38
23	703	23	297	265	30	735	562	7	438	37
24	726	24	274	295	31	705	569	7	431	36
25	750	23	250	326	30	674	576	7	424	35
26	773	23	227	356	30	644	583	7	417	34
27	796	24	204	386	30	614	590	6	410	33
28	820	23	180	416	30	584	596	7	404	32
29	843	23	157	446	30	554	603	7	397	31
30	67866	24	32134	73476	31	26524	05610	7	94390	30
31	890	23	110	507	30	493	617	7	383	29
32	913	23	087	537	30	463	624	7	376	28
33	936	23	064	567	30	433	631	7	369	27
34	959	23	041	597	30	403	638	7	362	26
35	982	24	018	627	30	373	645	6	355	25
36	68006	23	31994	657	30	343	651	7	349	24
37	029	23	971	687	30	313	658	7	342	23
38	052	23	948	717	30	283	665	7	335	22
39	075	23	925	747	30	253	672	7	328	21
40	098	23	902	777	30	223	679	7	321	20
41	121	23	879	807	30	193	686	7	314	19
42	144	23	856	837	30	163	693	7	307	18
43	167	23	833	867	30	133	700	7	300	17
44	190	23	810	897	30	103	707	7	293	16
45	213	24	787	927	30	073	714	7	286	15
46	237	23	763	957	30	043	721	6	279	14
47	260	23	740	987	30	013	727	7	273	13
48	283	22	717	74017	30	25983	734	7	266	12
49	305	23	695	047	30	953	741	7	259	11
50	328	23	672	077	30	923	748	7	252	10
51	351	23	649	107	30	893	755	7	245	9
52	374	23	626	137	29	863	762	7	238	8
53	397	23	603	166	30	834	769	7	231	7
54	420	23	580	196	30	804	776	7	224	6
55	443	23	557	226	30	774	783	7	217	5
56	466	23	534	256	30	744	790	7	210	4
57	489	23	511	286	30	714	797	7	203	3
58	512	22	488	316	29	684	804	7	196	2
59	534	23	466	345	30	655	811	7	189	1
60	68557		31443	74375		25625	05818		94182	0
′	9. l cos	d 1′	10. l sec	9. l cot	d 1′	10. l tan	10. l csc	d 1′	9. l sin	′

″	Proportional Parts							
	31	30	29	24	23	22	7	6
0	0	0	0	0	0	0	0	0
1	1	0	0	0	0	0	0	0
2	1	1	1	1	1	1	0	0
3	2	2	1	1	1	1	0	0
4	2	2	2	2	2	1	0	0
5	3	2	2	2	2	2	1	0
6	3	3	3	2	2	2	1	1
7	4	4	3	3	3	3	1	1
8	4	4	4	3	3	3	1	1
9	5	4	4	4	3	3	1	1
10	5	5	5	4	4	4	1	1
11	6	6	5	4	4	4	1	1
12	6	6	6	5	5	4	1	1
13	7	6	6	5	5	5	2	1
14	7	7	7	6	5	5	2	1
15	8	8	7	6	6	6	2	2
16	8	8	8	6	6	6	2	2
17	9	8	8	7	7	6	2	2
18	9	9	9	7	7	7	2	2
19	10	10	9	8	7	7	2	2
20	10	10	10	8	8	7	2	2
21	11	10	10	8	8	8	2	2
22	11	11	11	9	8	8	3	2
23	12	12	11	9	9	8	3	2
24	12	12	12	10	9	9	3	2
25	13	12	12	10	10	9	3	2
26	13	13	13	10	10	10	3	3
27	14	14	13	11	10	10	3	3
28	14	14	14	11	11	10	3	3
29	15	14	14	12	11	11	3	3
30	16	15	14	12	12	11	4	3
31	16	16	15	12	12	11	4	3
32	17	16	15	13	12	12	4	3
33	17	16	16	13	13	12	4	3
34	18	17	16	14	13	12	4	3
35	18	18	17	14	13	13	4	4
36	19	18	17	14	14	13	4	4
37	19	18	18	15	14	14	4	4
38	20	19	18	15	15	14	4	4
39	20	20	19	16	15	14	5	4
40	21	20	19	16	15	15	5	4
41	21	20	20	16	16	15	5	4
42	22	21	20	17	16	15	5	4
43	22	22	21	17	16	16	5	4
44	23	22	21	18	17	16	5	4
45	23	22	22	18	17	16	5	4
46	24	23	22	18	18	17	5	5
47	24	24	23	19	18	17	5	5
48	25	24	23	19	18	18	6	5
49	25	24	24	20	19	18	6	5
50	26	25	24	20	19	18	6	5
51	26	26	25	20	20	19	6	5
52	27	26	25	21	20	19	6	5
53	27	26	26	21	20	19	6	5
54	28	27	26	22	21	20	6	5
55	28	28	27	22	21	20	6	6
56	29	28	27	22	21	21	7	6
57	29	28	28	23	22	21	7	6
58	30	29	28	23	22	21	7	6
59	30	30	29	24	23	22	7	6
60	31	30	29	24	23	22	7	6
″	31	30	29	24	23	22	7	6
	Proportional Parts							

′	*l* sin 9.	d 1′	*l* csc 10.	*l* tan 9.	d 1′	*l* cot 10.	*l* sec 10.	d 1′	*l* cos 9.	′
0	68557	23	31443	74375	30	25625	05818	7	94182	60
1	580	23	420	405	30	595	825	7	175	59
2	603	22	397	435	30	565	832	7	168	58
3	625	23	375	465	29	535	839	7	161	57
4	648	23	352	494	30	506	846	7	154	56
5	671	23	329	524	30	476	853	7	147	55
6	694	22	306	554	29	446	860	7	140	54
7	716	23	284	583	30	417	867	7	133	53
8	739	23	261	613	30	387	874	7	126	52
9	762	22	238	643	30	357	881	7	119	51
10	784	23	216	673	29	327	888	7	112	50
11	807	22	193	702	30	298	895	7	105	49
12	829	23	171	732	30	268	902	8	098	48
13	852	23	148	762	29	238	910	7	090	47
14	875	22	125	791	30	209	917	7	083	46
15	897	23	103	821	30	179	924	7	076	45
16	920	22	080	851	29	149	931	7	069	44
17	942	23	058	880	30	120	938	7	062	43
18	965	22	035	910	29	090	945	7	055	42
19	987	23	013	939	30	061	952	7	048	41
20	69010	22	30990	969	29	031	959	7	041	40
21	032	23	968	998	30	002	966	7	034	39
22	055	22	945	75028	30	24972	973	7	027	38
23	077	23	923	058	29	942	980	8	020	37
24	100	22	900	087	30	913	988	7	012	36
25	122	22	878	117	29	883	995	7	005	35
26	144	23	856	146	30	854	06002	7	93998	34
27	167	22	833	176	29	824	009	7	991	33
28	189	23	811	205	30	795	016	7	984	32
29	212	22	788	235	29	765	023	7	977	31
30	69234	22	30766	75264	30	24736	06030	7	93970	30
31	256	23	744	294	29	706	037	8	963	29
32	279	22	721	323	30	677	045	7	955	28
33	301	22	699	353	29	647	052	7	948	27
34	323	22	677	382	29	618	059	7	941	26
35	345	23	655	411	30	589	066	7	934	25
36	368	22	632	441	29	559	073	7	927	24
37	390	22	610	470	30	530	080	8	920	23
38	412	22	588	500	29	500	088	7	912	22
39	434	22	566	529	29	471	095	7	905	21
40	456	23	544	558	30	442	102	7	898	20
41	479	22	521	588	29	412	109	7	891	19
42	501	22	499	617	30	383	116	8	884	18
43	523	22	477	647	29	353	124	7	876	17
44	545	22	455	676	29	324	131	7	869	16
45	567	22	433	705	30	295	138	7	862	15
46	589	22	411	735	29	265	145	8	855	14
47	611	22	389	764	29	236	153	7	847	13
48	633	22	367	793	29	207	160	7	840	12
49	655	22	345	822	30	178	167	7	833	11
50	677	22	323	852	29	148	174	7	826	10
51	699	22	301	881	29	119	181	8	819	9
52	721	22	279	910	29	090	189	7	811	8
53	743	22	257	939	30	061	196	7	804	7
54	765	22	235	969	29	031	203	8	797	6
55	787	22	213	998	29	002	211	7	789	5
56	809	22	191	76027	29	23973	218	7	782	4
57	831	22	169	056	30	944	225	7	775	3
58	853	22	147	086	29	914	232	8	768	2
59	875	22	125	115	29	885	240	7	760	1
60	69897		30103	76144		23856	06247		93753	0
′	9. *l* cos	d 1′	10. *l* sec	9. *l* cot	d 1′	10. *l* tan	10. *l* csc	d 1′	9. *l* sin	′

″	Proportional Parts					
	30	29	23	22	8	7
0	0	0	0	0	0	0
1	0	0	0	0	0	0
2	1	1	1	1	0	0
3	2	1	1	1	0	0
4	2	2	2	1	1	0
5	2	2	2	2	1	1
6	3	3	2	2	1	1
7	4	3	3	3	1	1
8	4	4	3	3	1	1
9	4	4	3	3	1	1
10	5	5	4	4	1	1
11	6	5	4	4	1	1
12	6	6	5	4	2	1
13	6	6	5	5	2	2
14	7	7	5	5	2	2
15	8	7	6	6	2	2
16	8	8	6	6	2	2
17	8	8	7	6	2	2
18	9	9	7	7	2	2
19	10	9	7	7	3	2
20	10	10	8	7	3	2
21	10	10	8	8	3	2
22	11	11	8	8	3	3
23	12	11	9	8	3	3
24	12	12	9	9	3	3
25	12	12	10	9	3	3
26	13	13	10	10	3	3
27	14	13	10	10	4	3
28	14	14	11	10	4	3
29	14	14	11	11	4	3
30	15	14	12	11	4	4
31	16	15	12	11	4	4
32	16	15	12	12	4	4
33	16	16	13	12	4	4
34	17	16	13	12	5	4
35	18	17	13	13	5	4
36	18	17	14	13	5	4
37	18	18	14	14	5	4
38	19	18	15	14	5	4
39	20	19	15	14	5	5
40	20	19	15	15	5	5
41	20	20	16	15	5	5
42	21	20	16	15	6	5
43	22	21	16	16	6	5
44	22	21	17	16	6	5
45	22	22	17	16	6	5
46	23	22	18	17	6	5
47	24	23	18	17	6	5
48	24	23	18	18	6	6
49	24	24	19	18	7	6
50	25	24	19	18	7	6
51	26	25	20	19	7	6
52	26	25	20	19	7	6
53	26	26	20	19	7	6
54	27	26	21	20	7	6
55	28	27	21	20	7	6
56	28	27	21	21	7	7
57	28	28	22	21	8	7
58	29	28	22	21	8	7
59	30	29	23	22	8	7
60	30	29	23	22	8	7
″	30	29	23	22	8	7
	Proportional Parts					

119° 60°

′	l sin 9.	d 1′	l csc 10.	l tan 9.	d 1′	l cot 10.	l sec 10.	d 1′	l cos 9.	′
0	69897	22	30103	76144	29	23856	06247	7	93753	60
1	919	22	081	173	29	827	254	8	746	59
2	941	22	059	202	29	798	262	7	738	58
3	963	21	037	231	30	769	269	7	731	57
4	984	22	016	261	29	739	276	7	724	56
5	70006	22	29994	290	29	710	283	8	717	55
6	028	22	972	319	29	681	291	7	709	54
7	050	22	950	348	29	652	298	7	702	53
8	072	21	928	377	29	623	305	8	695	52
9	093	22	907	406	29	594	313	7	687	51
10	115	22	885	435	29	565	320	7	680	50
11	137	22	863	464	29	536	327	8	673	49
12	159	21	841	493	29	507	335	7	665	48
13	180	22	820	522	29	478	342	8	658	47
14	202	22	798	551	29	449	350	7	650	46
15	224	21	776	580	29	420	357	7	643	45
16	245	22	755	609	30	391	364	8	636	44
17	267	21	733	639	29	361	372	7	628	43
18	288	22	712	668	29	332	379	7	621	42
19	310	22	690	697	28	303	386	8	614	41
20	332	21	668	725	29	275	394	7	606	40
21	353	22	647	754	29	246	401	8	599	39
22	375	21	625	783	29	217	409	7	591	38
23	396	22	604	812	29	188	416	7	584	37
24	418	21	582	841	29	159	423	8	577	36
25	439	22	561	870	29	130	431	7	569	35
26	461	21	539	899	29	101	438	8	562	34
27	482	22	518	928	29	072	446	7	554	33
28	504	21	496	957	29	043	453	8	547	32
29	525	22	475	986	29	014	461	7	539	31
30	70547	21	29453	77015	29	22985	06468	7	93532	30
31	568	22	432	044	29	956	475	8	525	29
32	590	21	410	073	28	927	483	7	517	28
33	611	22	389	101	29	899	490	8	510	27
34	633	21	367	130	29	870	498	7	502	26
35	654	21	346	159	29	841	505	8	495	25
36	675	22	325	188	29	812	513	7	487	24
37	697	21	303	217	29	783	520	8	480	23
38	718	21	282	246	28	754	528	7	472	22
39	739	22	261	274	29	726	535	8	465	21
40	761	21	239	303	29	697	543	7	457	20
41	782	21	218	332	29	668	550	8	450	19
42	803	21	197	361	29	639	558	7	442	18
43	824	22	176	390	28	610	565	8	435	17
44	846	21	154	418	29	582	573	7	427	16
45	867	21	133	447	29	553	580	8	420	15
46	888	21	112	476	29	524	588	7	412	14
47	909	22	091	505	28	495	595	8	405	13
48	931	21	069	533	29	467	603	7	397	12
49	952	21	048	562	29	438	610	8	390	11
50	973	21	027	591	28	409	618	7	382	10
51	994	21	006	619	29	381	625	8	375	9
52	71015	21	28985	648	29	352	633	7	367	8
53	036	22	964	677	29	323	640	8	360	7
54	058	21	942	706	28	294	648	8	352	6
55	079	21	921	734	29	266	656	7	344	5
56	100	21	900	763	28	237	663	8	337	4
57	121	21	879	791	29	209	671	7	329	3
58	142	21	858	820	29	180	678	8	322	2
59	163	21	837	849	28	151	686	7	314	1
60	71184		28816	77877		22123	06693		93307	0
′	9. l cos	d 1′	10. l sec	9. l cot	d 1′	10. l tan	10. l csc	d 1′	9. l sin	′

″	Proportional Parts 30	29	28	22	21	8	7
0	0	0	0	0	0	0	0
1	0	0	0	0	0	0	0
2	1	1	1	1	1	0	0
3	2	1	1	1	1	0	0
4	2	2	2	1	1	1	0
5	2	2	2	2	2	1	1
6	3	3	3	2	2	1	1
7	4	3	3	3	2	1	1
8	4	4	4	3	3	1	1
9	4	4	4	3	3	1	1
10	5	5	5	4	4	1	1
11	6	5	5	4	4	1	1
12	6	6	6	4	4	2	1
13	6	6	6	5	5	2	2
14	7	7	7	5	5	2	2
15	8	7	7	6	5	2	2
16	8	8	7	6	6	2	2
17	8	8	8	6	6	2	2
18	9	9	8	7	6	2	2
19	10	9	9	7	7	3	2
20	10	10	9	7	7	3	2
21	10	10	10	8	7	3	2
22	11	11	10	8	8	3	3
23	12	11	11	8	8	3	3
24	12	12	11	9	8	3	3
25	12	12	12	9	9	3	3
26	13	13	12	10	9	3	3
27	14	13	13	10	9	4	3
28	14	14	13	10	10	4	3
29	14	14	14	11	10	4	3
30	15	14	14	11	10	4	4
31	16	15	14	11	11	4	4
32	16	15	15	12	11	4	4
33	16	16	15	12	12	4	4
34	17	16	16	12	12	5	4
35	18	17	16	13	12	5	4
36	18	17	17	13	13	5	4
37	18	18	17	14	13	5	4
38	19	18	18	14	13	5	4
39	20	19	18	14	14	5	5
40	20	19	19	15	14	5	5
41	20	20	19	15	14	5	5
42	21	20	20	15	15	6	5
43	22	21	20	16	15	6	5
44	22	21	21	16	15	6	5
45	22	22	21	16	16	6	5
46	23	22	21	17	16	6	5
47	24	23	22	17	16	6	5
48	24	23	22	18	17	6	6
49	24	24	23	18	17	7	6
50	25	24	23	18	18	7	6
51	26	25	24	19	18	7	6
52	26	25	24	19	18	7	6
53	26	26	25	19	19	7	6
54	27	26	25	20	19	7	6
55	28	27	26	20	19	7	6
56	28	27	26	21	20	7	7
57	28	28	27	21	20	8	7
58	29	28	27	21	20	8	7
59	30	29	28	22	21	8	7
60	30	29	28	22	21	8	7
″	30 Proportional Parts	29	28	22	21	8	7

′	l sin 9.	d 1′	l csc 10.	l tan 9.	d 1′	l cot 10.	l sec 10.	d 1′	l cos 9.	′
0	**71**184	21	**28**816	**77**877	29	**22**123	**06**693	8	**93**307	**60**
1	205	21	795	906	29	094	701	8	299	59
2	226	21	774	935	28	065	709	7	291	58
3	247	21	753	963	29	037	716	8	284	57
4	268	21	732	992	28	008	724	7	276	56
5	289	21	711	**78**020	29	**21**980	731	8	269	**55**
6	310	21	690	049	28	951	739	8	261	54
7	331	21	669	077	29	923	747	7	253	53
8	352	21	648	106	29	894	754	8	246	52
9	373	20	627	135	28	865	762	8	238	51
10	393	21	607	163	29	837	770	7	230	**50**
11	414	21	586	192	28	808	777	8	223	49
12	435	21	565	220	29	780	785	8	215	48
13	456	21	544	249	28	751	793	7	207	47
14	477	21	523	277	29	723	800	8	200	46
15	498	21	502	306	28	694	808	8	192	**45**
16	519	20	481	334	29	666	816	7	184	44
17	539	21	461	363	28	637	823	8	177	43
18	560	21	440	391	28	609	831	8	169	42
19	581	21	419	419	29	581	839	7	161	41
20	602	20	398	448	28	552	846	8	154	**40**
21	622	21	378	476	29	524	854	8	146	39
22	643	21	357	505	28	495	862	7	138	38
23	664	21	336	533	29	467	869	8	131	37
24	685	20	315	562	28	438	877	8	123	36
25	705	21	295	590	28	410	885	7	115	**35**
26	726	21	274	618	29	382	892	8	108	34
27	747	20	253	647	28	353	900	8	100	33
28	767	21	233	675	29	325	908	8	092	32
29	788	21	212	704	28	296	916	7	084	31
30	**71**809	20	**28**191	**78**732	28	**21**268	**06**923	8	**93**077	**30**
31	829	21	171	760	29	240	931	8	069	29
32	850	20	150	789	28	211	939	8	061	28
33	870	21	130	817	28	183	947	7	053	27
34	891	20	109	845	29	155	954	8	046	26
35	911	21	089	874	28	126	962	8	038	**25**
36	932	20	068	902	28	098	970	8	030	24
37	952	21	048	930	29	070	978	8	022	23
38	973	21	027	959	28	041	986	7	014	22
39	994	20	006	987	28	013	993	8	007	21
40	**72**014	20	**27**986	**79**015	28	**20**985	**07**001	8	**92**999	**20**
41	034	21	966	043	29	957	009	8	991	19
42	055	20	945	072	28	928	017	7	983	18
43	075	21	925	100	28	900	024	8	976	17
44	096	20	904	128	28	872	032	8	968	16
45	116	21	884	156	29	844	040	8	960	**15**
46	137	20	863	185	28	815	048	8	952	14
47	157	20	843	213	28	787	056	8	944	13
48	177	21	823	241	28	759	064	7	936	12
49	198	20	802	269	28	731	071	8	929	11
50	218	20	782	297	29	703	079	8	921	**10**
51	238	21	762	326	28	674	087	8	913	9
52	259	20	741	354	28	646	095	8	905	8
53	279	20	721	382	28	618	103	8	897	7
54	299	21	701	410	28	590	111	8	889	6
55	320	20	680	438	28	562	119	7	881	**5**
56	340	20	660	466	29	534	126	8	874	4
57	360	21	640	495	28	505	134	8	866	3
58	381	20	619	523	28	477	142	8	858	2
59	401	20	599	551	28	449	150	8	850	1
60	**72**421		**27**579	**79**579		**20**421	**07**158		**92**842	**0**
′	9. l cos	d 1′	10. l sec	9. l cot	d 1′	10. l tan	10. l csc	d 1′	9. l sin	′

121° 58°

″	Proportional Parts 29	28	21	20	8	7
0	0	0	0	0	0	0
1	0	0	0	0	0	0
2	1	1	1	1	0	0
3	1	1	1	1	0	0
4	2	2	1	1	1	0
5	2	2	2	2	1	1
6	3	3	2	2	1	1
7	3	3	2	2	1	1
8	4	4	3	3	1	1
9	4	4	3	3	1	1
10	5	5	4	3	1	1
11	5	5	4	4	1	1
12	6	6	4	4	2	1
13	6	6	5	4	2	2
14	7	7	5	5	2	2
15	7	7	5	5	2	2
16	8	7	6	5	2	2
17	8	8	6	6	2	2
18	9	8	6	6	2	2
19	9	9	7	6	3	2
20	10	9	7	7	3	2
21	10	10	7	7	3	2
22	11	10	8	7	3	3
23	11	11	8	8	3	3
24	12	11	8	8	3	3
25	12	12	9	8	3	3
26	13	12	9	9	3	3
27	13	13	9	9	4	3
28	14	13	10	9	4	3
29	14	14	10	10	4	3
30	14	14	10	10	4	4
31	15	14	11	10	4	4
32	15	15	11	11	4	4
33	16	15	12	11	4	4
34	16	16	12	11	5	4
35	17	16	12	12	5	4
36	17	17	13	12	5	4
37	18	17	13	12	5	4
38	18	18	13	13	5	4
39	19	18	14	13	5	5
40	19	19	14	13	5	5
41	20	19	14	14	5	5
42	20	20	15	14	6	5
43	21	20	15	14	6	5
44	21	21	15	15	6	5
45	22	21	16	15	6	5
46	22	21	16	15	6	5
47	23	22	16	16	6	5
48	23	22	17	16	6	6
49	24	23	17	16	7	6
50	24	23	18	17	7	6
51	25	24	18	17	7	6
52	25	24	18	17	7	6
53	26	25	19	18	7	6
54	26	25	19	18	7	6
55	27	26	19	18	7	6
56	27	26	20	19	7	7
57	28	27	20	19	8	7
58	28	27	20	19	8	7
59	29	28	21	20	8	7
60	**29**	**28**	**21**	**20**	**8**	**7**
″	Proportional Parts 29	28	21	20	8	7

′	*l* sin 9.	d 1′	*l* csc 10.	*l* tan 9.	d 1′	*l* cot 10.	*l* sec 10.	d 1′	*l* cos 9.	′
0	72421	20	27579	79579	28	20421	07158	8	92842	60
1	441	20	559	607	28	393	166	8	834	59
2	461	21	539	635	28	365	174	8	826	58
3	482	20	518	663	28	337	182	8	818	57
4	502	20	498	691	28	309	190	7	810	56
5	522	20	478	719	28	281	197	8	803	55
6	542	20	458	747	29	253	205	8	795	54
7	562	20	438	776	28	224	213	8	787	53
8	582	20	418	804	28	196	221	8	779	52
9	602	20	398	832	28	168	229	8	771	51
10	622	21	378	860	28	140	237	8	763	50
11	643	20	357	888	28	112	245	8	755	49
12	663	20	337	916	28	084	253	8	747	48
13	683	20	317	944	28	056	261	8	739	47
14	703	20	297	972	28	028	269	8	731	46
15	723	20	277	80000	28	000	277	8	723	45
16	743	20	257	028	28	19972	285	8	715	44
17	763	20	237	056	28	944	293	8	707	43
18	783	20	217	084	28	916	301	8	699	42
19	803	20	197	112	28	888	309	8	691	41
20	823	20	177	140	28	860	317	8	683	40
21	843	20	157	168	27	832	325	8	675	39
22	863	20	137	195	28	805	333	8	667	38
23	883	19	117	223	28	777	341	8	659	37
24	902	20	098	251	28	749	349	8	651	36
25	922	20	078	279	28	721	357	8	643	35
26	942	20	058	307	28	693	365	8	635	34
27	962	20	038	335	28	665	373	8	627	33
28	982	20	018	363	28	637	381	8	619	32
29	73002	20	26998	391	28	609	389	8	611	31
30	73022	19	26978	80419	28	19581	07397	8	92603	30
31	041	20	959	447	27	553	405	8	595	29
32	061	20	939	474	28	526	413	8	587	28
33	081	20	919	502	28	498	421	8	579	27
34	101	20	899	530	28	470	429	8	571	26
35	121	19	879	558	28	442	437	8	563	25
36	140	20	860	586	28	414	445	9	555	24
37	160	20	840	614	28	386	454	8	546	23
38	180	20	820	642	27	358	462	8	538	22
39	200	19	800	669	28	331	470	8	530	21
40	219	20	781	697	28	303	478	8	522	20
41	239	20	761	725	28	275	486	8	514	19
42	259	19	741	753	28	247	494	8	506	18
43	278	20	722	781	27	219	502	8	498	17
44	298	20	702	808	28	192	510	8	490	16
45	318	19	682	836	28	164	518	9	482	15
46	337	20	663	864	28	136	527	8	473	14
47	357	20	643	892	27	108	535	8	465	13
48	377	19	623	919	28	081	543	8	457	12
49	396	20	604	947	28	053	551	8	449	11
50	416	19	584	975	28	025	559	8	441	10
51	435	20	565	81003	27	18997	567	8	433	9
52	455	19	545	030	28	970	575	9	425	8
53	474	20	526	058	28	942	584	8	416	7
54	494	19	506	086	27	914	592	8	408	6
55	513	20	487	113	28	887	600	8	400	5
56	533	19	467	141	28	859	608	8	392	4
57	552	20	448	169	27	831	616	8	384	3
58	572	19	428	196	28	804	624	9	376	2
59	591	20	409	224	28	776	633	8	367	1
60	73611		26389	81252		18748	07641		92359	0
′	9. *l* cos	d 1′	10. *l* sec	9. *l* cot	d 1′	10. *l* tan	10. *l* csc	d 1′	9. *l* sin	′

″	Proportional Parts 29	28	27	21	20	19	9	8	7
0	0	0	0	0	0	0	0	0	0
1	0	0	0	0	0	0	0	0	0
2	1	1	1	1	1	1	0	0	0
3	1	1	1	1	1	1	0	0	0
4	2	2	2	1	1	1	1	1	0
5	2	2	2	2	2	2	1	1	1
6	3	3	3	2	2	2	1	1	1
7	3	3	3	2	2	2	1	1	1
8	4	4	4	3	3	3	1	1	1
9	4	4	4	3	3	3	1	1	1
10	5	5	4	4	3	3	2	1	1
11	5	5	5	4	4	3	2	1	1
12	6	6	5	4	4	4	2	2	1
13	6	6	6	5	4	4	2	2	2
14	7	7	6	5	5	4	2	2	2
15	7	7	7	5	5	5	2	2	2
16	8	7	7	6	5	5	2	2	2
17	8	8	8	6	6	5	3	2	2
18	9	8	8	6	6	6	3	2	2
19	9	9	9	7	6	6	3	3	2
20	10	9	9	7	7	6	3	3	2
21	10	10	9	7	7	7	3	3	2
22	11	10	10	8	7	7	3	3	3
23	11	11	10	8	8	7	3	3	3
24	12	11	11	8	8	8	4	3	3
25	12	12	11	9	8	8	4	3	3
26	13	12	12	9	9	8	4	3	3
27	13	13	12	9	9	9	4	4	3
28	14	13	13	10	9	9	4	4	3
29	14	14	13	10	10	9	4	4	3
30	14	14	14	10	10	10	4	4	4
31	15	14	14	11	10	10	5	4	4
32	15	15	14	11	11	10	5	4	4
33	16	15	15	12	11	10	5	4	4
34	16	16	15	12	11	11	5	5	4
35	17	16	16	12	12	11	5	5	4
36	17	17	16	13	12	11	5	5	4
37	18	17	17	13	12	12	6	5	4
38	18	18	17	13	13	12	6	5	4
39	19	18	18	14	13	12	6	5	5
40	19	19	18	14	13	13	6	5	5
41	20	19	18	14	14	13	6	5	5
42	20	20	19	15	14	13	6	6	5
43	21	20	19	15	14	14	6	6	5
44	21	21	20	15	15	14	7	6	5
45	22	21	20	16	15	14	7	6	5
46	22	21	21	16	15	15	7	6	5
47	23	22	21	16	16	15	7	6	5
48	23	22	22	17	16	15	7	6	6
49	24	23	22	17	16	16	7	7	6
50	24	23	22	18	17	16	8	7	6
51	25	24	23	18	17	16	8	7	6
52	25	24	23	18	17	16	8	7	6
53	26	25	24	19	18	17	8	7	6
54	26	25	24	19	18	17	8	7	6
55	27	26	25	19	18	17	8	7	6
56	27	26	25	20	19	18	8	7	7
57	28	27	26	20	19	18	9	8	7
58	28	27	26	20	19	18	9	8	7
59	29	28	27	21	20	19	9	8	7
60	29	28	27	21	20	19	9	8	7
″	29 Proportional Parts	28	27	21	20	19	9	8	7

′	*l* sin 9.	d 1′	*l* csc 10.	*l* tan 9.	d 1′	*l* cot 10.	*l* sec 10.	d 1′	*l* cos 9.	′
0	73611	19	26389	81252	27	18748	07641	8	92359	60
1	630	20	370	279	28	721	649	8	351	59
2	650	19	350	307	28	693	657	8	343	58
3	669	20	331	335	27	665	665	9	335	57
4	689	19	311	362	28	638	674	8	326	56
5	708	19	292	390	28	610	682	8	318	55
6	727	20	273	418	27	582	690	8	310	54
7	747	19	253	445	28	555	698	9	302	53
8	766	19	234	473	27	527	707	8	293	52
9	785	20	215	500	28	500	715	8	285	51
10	805	19	195	528	28	472	723	8	277	50
11	824	19	176	556	27	444	731	9	269	49
12	843	20	157	583	28	417	740	8	260	48
13	863	19	137	611	27	389	748	8	252	47
14	882	19	118	638	28	362	756	9	244	46
15	901	20	099	666	27	334	765	8	235	45
16	921	19	079	693	28	307	773	8	227	44
17	940	19	060	721	27	279	781	8	219	43
18	959	19	041	748	28	252	789	9	211	42
19	978	19	022	776	27	224	798	8	202	41
20	997	20	003	803	28	197	806	8	194	40
21	74017	19	25983	831	27	169	814	9	186	39
22	036	19	964	858	28	142	823	8	177	38
23	055	19	945	886	27	114	831	8	169	37
24	074	19	926	913	28	087	839	9	161	36
25	093	20	907	941	27	059	848	8	152	35
26	113	19	887	968	28	032	856	8	144	34
27	132	19	868	996	27	004	864	9	136	33
28	151	19	849	82023	28	17977	873	8	127	32
29	170	19	830	051	27	949	881	8	119	31
30	74189	19	25811	82078	28	17922	07889	9	92111	30
31	208	19	792	106	27	894	898	8	102	29
32	227	19	773	133	28	867	906	8	094	28
33	246	19	754	161	27	839	914	9	086	27
34	265	19	735	188	27	812	923	8	077	26
35	284	19	716	215	28	785	931	9	069	25
36	303	19	697	243	27	757	940	8	060	24
37	322	19	678	270	28	730	948	8	052	23
38	341	19	659	298	27	702	956	9	044	22
39	360	19	640	325	27	675	965	8	035	21
40	379	19	621	352	28	648	973	9	027	20
41	398	19	602	380	27	620	982	8	018	19
42	417	19	583	407	28	593	990	8	010	18
43	436	19	564	435	27	565	998	9	002	17
44	455	19	545	462	27	538	08007	8	91993	16
45	474	19	526	489	28	511	015	9	985	15
46	493	19	507	517	27	483	024	8	976	14
47	512	19	488	544	27	456	032	9	968	13
48	531	18	469	571	28	429	041	8	959	12
49	549	19	451	599	27	401	049	9	951	11
50	568	19	432	626	27	374	058	8	942	10
51	587	19	413	653	28	347	066	9	934	9
52	606	19	394	681	27	319	075	8	925	8
53	625	19	375	708	27	292	083	9	917	7
54	644	18	356	735	27	265	092	8	908	6
55	662	19	338	762	28	238	100	9	900	5
56	681	19	319	790	27	210	109	8	891	4
57	700	19	300	817	27	183	117	9	883	3
58	719	18	281	844	27	156	126	8	874	2
59	737	19	263	871	28	129	134	9	866	1
60	74756		25244	82899		17101	08143		91857	0
′	9. *l* cos	d 1′	10. *l* sec	9. *l* cot	d 1′	10. *l* tan	10. *l* csc	d 1′	9. *l* sin	′

″	Proportional Parts						
	28	27	20	19	18	9	8
0	0	0	0	0	0	0	0
1	0	0	0	0	0	0	0
2	1	1	1	1	1	0	0
3	1	1	1	1	1	0	0
4	2	2	1	1	1	1	1
5	2	2	2	2	2	1	1
6	3	3	2	2	2	1	1
7	3	3	2	2	2	1	1
8	4	4	3	3	2	1	1
9	4	4	3	3	3	1	1
10	5	4	3	3	3	2	1
11	5	5	4	3	3	2	1
12	6	5	4	4	4	2	2
13	6	6	4	4	4	2	2
14	7	6	5	4	4	2	2
15	7	7	5	5	4	2	2
16	7	7	5	5	5	2	2
17	8	8	6	5	5	3	2
18	8	8	6	6	5	3	2
19	9	9	6	6	6	3	3
20	9	9	7	6	6	3	3
21	10	9	7	7	6	3	3
22	10	10	7	7	7	3	3
23	11	10	8	7	7	3	3
24	11	11	8	8	7	4	3
25	12	11	8	8	8	4	3
26	12	12	9	8	8	4	3
27	13	12	9	9	8	4	4
28	13	13	9	9	8	4	4
29	14	13	10	9	9	4	4
30	14	14	10	10	9	4	4
31	14	14	10	10	9	5	4
32	15	14	11	10	10	5	4
33	15	15	11	10	10	5	4
34	16	15	11	11	10	5	5
35	16	16	12	11	10	5	5
36	17	16	12	11	11	5	5
37	17	17	12	12	11	6	5
38	18	17	13	12	11	6	5
39	18	18	13	12	12	6	5
40	19	18	13	13	12	6	5
41	19	18	14	13	12	6	5
42	20	19	14	13	13	6	6
43	20	19	14	14	13	6	6
44	21	20	15	14	13	7	6
45	21	20	15	14	14	7	6
46	21	21	15	15	14	7	6
47	22	21	16	15	14	7	6
48	22	22	16	15	14	7	6
49	23	22	16	16	15	7	7
50	23	22	17	16	15	8	7
51	24	23	17	16	15	8	7
52	24	23	17	16	16	8	7
53	25	24	18	17	16	8	7
54	25	24	18	17	16	8	7
55	26	25	18	17	16	8	7
56	26	25	19	18	17	8	7
57	27	26	19	18	17	9	8
58	27	26	19	18	17	9	8
59	28	27	20	19	18	9	8
60	28	27	20	19	18	9	8
″	28	27	20	19	18	9	8
	Proportional Parts						

′	l sin 9.	d 1′	l csc 10.	l tan 9.	d 1′	l cot 10.	l sec 10.	d 1′	l cos 9.	′
0	74756	19	25244	82899	27	17101	08143	8	91857	60
1	775	19	225	926	27	074	151	9	849	59
2	794	18	206	953	27	047	160	8	840	58
3	812	19	188	980	28	020	168	9	832	57
4	831	19	169	83008	27	16992	177	8	823	56
5	850	18	150	035	27	965	185	9	815	55
6	868	19	132	062	27	938	194	8	806	54
7	887	19	113	089	28	911	202	9	798	53
8	906	18	094	117	27	883	211	8	789	52
9	924	19	076	144	27	856	219	9	781	51
10	943	18	057	171	27	829	228	9	772	50
11	961	19	039	198	27	802	237	8	763	49
12	980	19	020	225	27	775	245	9	755	48
13	999	18	001	252	28	748	254	8	746	47
14	75017	19	24983	280	27	720	262	9	738	46
15	036	18	964	307	27	693	271	9	729	45
16	054	19	946	334	27	666	280	8	720	44
17	073	18	927	361	27	639	288	9	712	43
18	091	19	909	388	27	612	297	8	703	42
19	110	18	890	415	27	585	305	9	695	41
20	128	19	872	442	28	558	314	9	686	40
21	147	18	853	470	27	530	323	8	677	39
22	165	19	835	497	27	503	331	9	669	38
23	184	18	816	524	27	476	340	9	660	37
24	202	19	798	551	27	449	349	8	651	36
25	221	18	779	578	27	422	357	9	643	35
26	239	19	761	605	27	395	366	9	634	34
27	258	18	742	632	27	368	375	8	625	33
28	276	18	724	659	27	341	383	9	617	32
29	294	19	706	686	27	314	392	9	608	31
30	75313	18	24687	83713	27	16287	08401	8	91599	30
31	331	19	669	740	28	260	409	9	591	29
32	350	18	650	768	27	232	418	9	582	28
33	368	18	632	795	27	205	427	8	573	27
34	386	19	614	822	27	178	435	9	565	26
35	405	18	595	849	27	151	444	9	556	25
36	423	18	577	876	27	124	453	9	547	24
37	441	18	559	903	27	097	462	8	538	23
38	459	19	541	930	27	070	470	9	530	22
39	478	18	522	957	27	043	479	9	521	21
40	496	18	504	984	27	016	488	8	512	20
41	514	19	486	84011	27	15989	496	9	504	19
42	533	18	467	038	27	962	505	9	495	18
43	551	18	449	065	27	935	514	9	486	17
44	569	18	431	092	27	908	523	8	477	16
45	587	18	413	119	27	881	531	9	469	15
46	605	19	395	146	27	854	540	9	460	14
47	624	18	376	173	27	827	549	9	451	13
48	642	18	358	200	27	800	558	9	442	12
49	660	18	340	227	27	773	567	8	433	11
50	678	18	322	254	26	746	575	9	425	10
51	696	18	304	280	27	720	584	9	416	9
52	714	19	286	307	27	693	593	9	407	8
53	733	18	267	334	27	666	602	9	398	7
54	751	18	249	361	27	639	611	8	389	6
55	769	18	231	388	27	612	619	9	381	5
56	787	18	213	415	27	585	628	9	372	4
57	805	18	195	442	27	558	637	9	363	3
58	823	18	177	469	27	531	646	9	354	2
59	841	18	159	496	27	504	655	9	345	1
60	75859		24141	84523		15477	08664		91336	0
′	9. l cos	d 1′	10. l sec	9. l cot	d 1′	10. l tan	10. l csc	d 1′	9. l sin	′

″	Proportional Parts 28	27	26	19	18	9	8
0	0	0	0	0	0	0	0
1	0	0	0	0	0	0	0
2	1	1	1	1	1	0	0
3	1	1	1	1	1	0	0
4	2	2	2	1	1	1	1
5	2	2	2	2	2	1	1
6	3	3	3	2	2	1	1
7	3	3	3	2	2	1	1
8	4	4	3	3	2	1	1
9	4	4	4	3	3	1	1
10	5	4	4	3	3	2	1
11	5	5	5	3	3	2	1
12	6	5	5	4	4	2	2
13	6	6	6	4	4	2	2
14	7	6	6	4	4	2	2
15	7	7	6	5	4	2	2
16	7	7	7	5	5	2	2
17	8	8	7	5	5	3	2
18	8	8	8	6	5	3	2
19	9	9	8	6	6	3	3
20	9	9	9	6	6	3	3
21	10	9	9	7	6	3	3
22	10	10	10	7	7	3	3
23	11	10	10	7	7	3	3
24	11	11	10	8	7	4	3
25	12	11	11	8	8	4	3
26	12	12	11	8	8	4	3
27	13	12	12	9	8	4	4
28	13	13	12	9	8	4	4
29	14	13	13	9	9	4	4
30	14	14	13	10	9	4	4
31	14	14	13	10	9	5	4
32	15	14	14	10	10	5	4
33	15	15	14	10	10	5	4
34	16	15	15	11	10	5	5
35	16	16	15	11	10	5	5
36	17	16	16	11	11	5	5
37	17	17	16	12	11	6	5
38	18	17	16	12	11	6	5
39	18	18	17	12	12	6	5
40	19	18	17	13	12	6	5
41	19	18	18	13	12	6	5
42	20	19	18	13	13	6	6
43	20	19	19	14	13	6	6
44	21	20	19	14	13	7	6
45	21	20	20	14	14	7	6
46	21	21	20	15	14	7	6
47	22	21	20	15	14	7	6
48	22	22	21	15	14	7	6
49	23	22	21	16	15	7	7
50	23	22	22	16	15	8	7
51	24	23	22	16	15	8	7
52	24	23	23	16	16	8	7
53	25	24	23	17	16	8	7
54	25	24	23	17	16	8	7
55	26	25	24	17	16	8	7
56	26	25	24	18	17	8	7
57	27	26	25	18	17	9	8
58	27	26	25	18	17	9	8
59	28	27	26	19	18	9	8
60	28	27	26	19	18	9	8
″	28	27	26	19	18	9	8

Proportional Parts

124° 55°

′	*l* sin 9.	d 1′	*l* csc 10.	*l* tan 9.	d 1′	*l* cot 10.	*l* sec 10.	d 1′	*l* cos 9.	′
0	75859	18	24141	84523	27	15477	08664	8	91336	60
1	877	18	123	550	26	450	672	9	328	59
2	895	18	105	576	27	424	681	9	319	58
3	913	18	087	603	27	397	690	9	310	57
4	931	18	069	630	27	370	699	9	301	56
5	949	18	051	657	27	343	708	9	292	55
6	967	18	033	684	27	316	717	9	283	54
7	985	18	015	711	27	289	726	8	274	53
8	76003	18	23997	738	26	262	734	9	266	52
9	021	18	979	764	27	236	743	9	257	51
10	039	18	961	791	27	209	752	9	248	50
11	057	18	943	818	27	182	761	9	239	49
12	075	18	925	845	27	155	770	9	230	48
13	093	18	907	872	27	128	779	9	221	47
14	111	18	889	899	26	101	788	9	212	46
15	129	17	871	925	27	075	797	9	203	45
16	146	18	854	952	27	048	806	9	194	44
17	164	18	836	979	27	021	815	9	185	43
18	182	18	818	85006	27	14994	824	9	176	42
19	200	18	800	033	26	967	833	9	167	41
20	218	18	782	059	27	941	842	9	158	40
21	236	17	764	086	27	914	851	8	149	39
22	253	18	747	113	27	887	859	9	141	38
23	271	18	729	140	26	860	868	9	132	37
24	289	18	711	166	27	834	877	9	123	36
25	307	17	693	193	27	807	886	9	114	35
26	324	18	676	220	27	780	895	9	105	34
27	342	18	658	247	26	753	904	9	096	33
28	360	18	640	273	27	727	913	9	087	32
29	378	17	622	300	27	700	922	9	078	31
30	76395	18	23605	85327	27	14673	08931	9	91069	30
31	413	18	587	354	26	646	940	9	060	29
32	431	17	569	380	27	620	949	9	051	28
33	448	18	552	407	27	593	958	9	042	27
34	466	18	534	434	26	566	967	10	033	26
35	484	17	516	460	27	540	977	9	023	25
36	501	18	499	487	27	513	986	9	014	24
37	519	18	481	514	26	486	995	9	005	23
38	537	17	463	540	27	460	09004	9	90996	22
39	554	18	446	567	27	433	013	9	987	21
40	572	18	428	594	26	406	022	9	978	20
41	590	17	410	620	27	380	031	9	969	19
42	607	18	393	647	27	353	040	9	960	18
43	625	17	375	674	26	326	049	9	951	17
44	642	18	358	700	27	300	058	9	942	16
45	660	17	340	727	27	273	067	9	933	15
46	677	18	323	754	26	246	076	9	924	14
47	695	17	305	780	27	220	085	9	915	13
48	712	18	288	807	27	193	094	10	906	12
49	730	17	270	834	26	166	104	9	896	11
50	747	18	253	860	27	140	113	9	887	10
51	765	17	235	887	26	113	122	9	878	9
52	782	18	218	913	27	087	131	9	869	8
53	800	17	200	940	27	060	140	9	860	7
54	817	18	183	967	26	033	149	9	851	6
55	835	17	165	993	27	007	158	10	842	5
56	852	18	148	86020	26	13980	168	9	832	4
57	870	17	130	046	27	954	177	9	823	3
58	887	17	113	073	27	927	186	9	814	2
59	904	18	096	100	26	900	195	9	805	1
60	76922		23078	86126		13874	09204		90796	0
′	9. *l* cos	d 1′	10. *l* sec	9. *l* cot	d 1′	10. *l* tan	10. *l* csc	d 1′	9. *l* sin	′

″	Proportional Parts 27	26	18	17	10	9	8
0	0	0	0	0	0	0	0
1	0	0	0	0	0	0	0
2	1	1	1	1	0	0	0
3	1	1	1	1	0	0	0
4	2	2	1	1	1	1	1
5	2	2	2	1	1	1	1
6	3	3	2	2	1	1	1
7	3	3	2	2	1	1	1
8	4	3	2	2	1	1	1
9	4	4	3	3	2	1	1
10	4	4	3	3	2	2	1
11	5	5	3	3	2	2	1
12	5	5	4	3	2	2	2
13	6	6	4	4	2	2	2
14	6	6	4	4	2	2	2
15	7	6	4	4	2	2	2
16	7	7	5	5	3	2	2
17	8	7	5	5	3	3	2
18	8	8	5	5	3	3	2
19	9	8	6	5	3	3	3
20	9	9	6	6	3	3	3
21	9	9	6	6	4	3	3
22	10	10	7	6	4	3	3
23	10	10	7	7	4	3	3
24	11	10	7	7	4	4	3
25	11	11	8	7	4	4	3
26	12	11	8	7	4	4	3
27	12	12	8	8	4	4	4
28	13	12	8	8	5	4	4
29	13	13	9	8	5	4	4
30	14	13	9	8	5	4	4
31	14	13	9	9	5	5	4
32	14	14	10	9	5	5	4
33	15	14	10	9	6	5	4
34	15	15	10	10	6	5	5
35	16	15	10	10	6	5	5
36	16	16	11	10	6	5	5
37	17	16	11	10	6	6	5
38	17	16	11	11	6	6	5
39	18	17	12	11	6	6	5
40	18	17	12	11	7	6	5
41	18	18	12	12	7	6	5
42	19	18	13	12	7	6	6
43	19	19	13	12	7	6	6
44	20	19	13	12	7	7	6
45	20	20	14	13	8	7	6
46	21	20	14	13	8	7	6
47	21	20	14	13	8	7	6
48	22	21	14	14	8	7	6
49	22	21	15	14	8	7	7
50	22	22	15	14	8	8	7
51	23	22	15	14	8	8	7
52	23	23	16	15	9	8	7
53	24	23	16	15	9	8	7
54	24	23	16	15	9	8	7
55	25	24	16	16	9	8	7
56	25	24	17	16	9	8	7
57	26	25	17	16	10	9	8
58	26	25	17	16	10	9	8
59	27	26	18	17	10	9	8
60	27	26	18	17	10	9	8
″	27	26	18	17	10	9	8

Proportional Parts

′	l sin 9.	d 1′	l csc 10.	l tan 9.	d 1′	l cot 10.	l sec 10.	d 1′	l cos 9.	′
0	76922	17	23078	86126	27	13874	09204	9	90796	60
1	939	18	061	153	26	847	213	10	787	59
2	957	17	043	179	27	821	223	9	777	58
3	974	17	026	206	26	794	232	9	768	57
4	991	18	009	232	27	768	241	9	759	56
5	77009	17	22991	259	26	741	250	9	750	55
6	026	17	974	285	27	715	259	10	741	54
7	043	18	957	312	26	688	269	9	731	53
8	061	17	939	338	27	662	278	9	722	52
9	078	17	922	365	27	635	287	9	713	51
10	095	17	905	392	26	608	296	10	704	50
11	112	18	888	418	27	582	306	9	694	49
12	130	17	870	445	26	555	315	9	685	48
13	147	17	853	471	27	529	324	9	676	47
14	164	17	836	498	26	502	333	10	667	46
15	181	18	819	524	27	476	343	9	657	45
16	199	17	801	551	26	449	352	9	648	44
17	216	17	784	577	26	423	361	9	639	43
18	233	17	767	603	27	397	370	10	630	42
19	250	18	750	630	26	370	380	9	620	41
20	268	17	732	656	27	344	389	9	611	40
21	285	17	715	683	26	317	398	10	602	39
22	302	17	698	709	27	291	408	9	592	38
23	319	17	681	736	26	264	417	9	583	37
24	336	17	664	762	27	238	426	9	574	36
25	353	17	647	789	26	211	435	10	565	35
26	370	17	630	815	27	185	445	9	555	34
27	387	18	613	842	26	158	454	9	546	33
28	405	17	595	868	26	132	463	10	537	32
29	422	17	578	894	27	106	473	9	527	31
30	77439	17	22561	86921	26	13079	09482	9	90518	30
31	456	17	544	947	27	053	491	10	509	29
32	473	17	527	974	26	026	501	9	499	28
33	490	17	510	87000	27	000	510	10	490	27
34	507	17	493	027	26	12973	520	9	480	26
35	524	17	476	053	26	947	529	9	471	25
36	541	17	459	079	27	921	538	10	462	24
37	558	17	442	106	26	894	548	9	452	23
38	575	17	425	132	26	868	557	9	443	22
39	592	17	408	158	27	842	566	10	434	21
40	609	17	391	185	26	815	576	9	424	20
41	626	17	374	211	27	789	585	10	415	19
42	643	17	357	238	26	762	595	9	405	18
43	660	17	340	264	26	736	604	10	396	17
44	677	17	323	290	27	710	614	9	386	16
45	694	17	306	317	26	683	623	9	377	15
46	711	17	289	343	26	657	632	10	368	14
47	728	16	272	369	27	631	642	9	358	13
48	744	17	256	396	26	604	651	10	349	12
49	761	17	239	422	26	578	661	9	339	11
50	778	17	222	448	27	552	670	10	330	10
51	795	17	205	475	26	525	680	9	320	9
52	812	17	188	501	26	499	689	10	311	8
53	829	17	171	527	27	473	699	9	301	7
54	846	16	154	554	26	446	708	10	292	6
55	862	17	138	580	26	420	718	9	282	5
56	879	17	121	606	27	394	727	10	273	4
57	896	17	104	633	26	367	737	9	263	3
58	913	17	087	659	26	341	746	10	254	2
59	930	16	070	685	26	315	756	9	244	1
60	77946		22054	87711		12289	09765		90235	0
′	9. l cos	d 1′	10. l sec	9. l cot	d 1′	10. l tan	10. l csc	d 1′	9. l sin	′

Proportional Parts

″	27	26	18	17	16	10	9
0	0	0	0	0	0	0	0
1	0	0	0	0	0	0	0
2	1	1	1	1	1	0	0
3	1	1	1	1	1	0	0
4	2	2	1	1	1	1	1
5	2	2	2	1	1	1	1
6	3	3	2	2	2	1	1
7	3	3	2	2	2	1	1
8	4	3	2	2	2	1	1
9	4	4	3	3	2	2	1
10	4	4	3	3	3	2	2
11	5	5	3	3	3	2	2
12	5	5	4	3	3	2	2
13	6	6	4	4	3	2	2
14	6	6	4	4	4	2	2
15	7	6	4	4	4	2	2
16	7	7	5	5	4	3	2
17	8	7	5	5	5	3	3
18	8	8	5	5	5	3	3
19	9	8	6	5	5	3	3
20	9	9	6	6	5	3	3
21	9	9	6	6	6	4	3
22	10	10	7	6	6	4	3
23	10	10	7	7	6	4	3
24	11	10	7	7	6	4	4
25	11	11	8	7	7	4	4
26	12	11	8	7	7	4	4
27	12	12	8	8	7	4	4
28	13	12	8	8	7	5	4
29	13	13	9	8	8	5	4
30	14	13	9	8	8	5	4
31	14	13	9	9	8	5	5
32	14	14	10	9	9	5	5
33	15	14	10	9	9	6	5
34	15	15	10	10	9	6	5
35	16	15	10	10	9	6	5
36	16	16	11	10	10	6	5
37	17	16	11	10	10	6	6
38	17	16	11	11	10	6	6
39	18	17	12	11	10	6	6
40	18	17	12	11	11	7	6
41	18	18	12	12	11	7	6
42	19	18	13	12	11	7	6
43	19	19	13	12	11	7	6
44	20	19	13	12	12	7	7
45	20	20	14	13	12	8	7
46	21	20	14	13	12	8	7
47	21	20	14	13	13	8	7
48	22	21	14	14	13	8	7
49	22	21	15	14	13	8	7
50	22	22	15	14	13	8	8
51	23	22	15	14	14	8	8
52	23	23	16	15	14	9	8
53	24	23	16	15	14	9	8
54	24	23	16	15	14	9	8
55	25	24	16	16	15	9	8
56	25	24	17	16	15	9	8
57	26	25	17	16	15	10	9
58	26	25	17	16	15	10	9
59	27	26	18	17	16	10	9
60	27	26	18	17	16	10	9
″	27	26	18	17	16	10	9

Proportional Parts

′	l sin 9.	d 1′	l csc 10.	l tan 9.	d 1′	l cot 10.	l sec 10.	d 1′	l cos 9.	′
0	77946	17	22054	87711	27	12289	09765	10	90235	60
1	963	17	037	738	26	262	775	9	225	59
2	980	17	020	764	26	236	784	10	216	58
3	997	16	003	790	27	210	794	9	206	57
4	78013	17	21987	817	26	183	803	10	197	56
5	030	17	970	843	26	157	813	9	187	55
6	047	16	953	869	26	131	822	10	178	54
7	063	17	937	895	27	105	832	9	168	53
8	080	17	920	922	26	078	841	10	159	52
9	097	16	903	948	26	052	851	10	149	51
10	113	17	887	974	26	026	861	9	139	50
11	130	17	870	88000	27	000	870	10	130	49
12	147	16	853	027	26	11973	880	9	120	48
13	163	17	837	053	26	947	889	10	111	47
14	180	17	820	079	26	921	899	10	101	46
15	197	16	803	105	26	895	909	9	091	45
16	213	17	787	131	27	869	918	10	082	44
17	230	16	770	158	26	842	928	9	072	43
18	246	17	754	184	26	816	937	10	063	42
19	263	17	737	210	26	790	947	10	053	41
20	280	16	720	236	26	764	957	9	043	40
21	296	17	704	262	27	738	966	10	034	39
22	313	16	687	289	26	711	976	10	024	38
23	329	17	671	315	26	685	986	9	014	37
24	346	16	654	341	26	659	995	10	005	36
25	362	17	638	367	26	633	10005	10	89995	35
26	379	16	621	393	27	607	015	9	985	34
27	395	17	605	420	26	580	024	10	976	33
28	412	16	588	446	26	554	034	10	966	32
29	428	17	572	472	26	528	044	9	956	31
30	78445	16	21555	88498	26	11502	10053	10	89947	30
31	461	17	539	524	26	476	063	10	937	29
32	478	16	522	550	27	450	073	9	927	28
33	494	16	506	577	26	423	082	10	918	27
34	510	17	490	603	26	397	092	10	908	26
35	527	16	473	629	26	371	102	10	898	25
36	543	17	457	655	26	345	112	9	888	24
37	560	16	440	681	26	319	121	10	879	23
38	576	16	424	707	26	293	131	10	869	22
39	592	17	408	733	26	267	141	10	859	21
40	609	16	391	759	27	241	151	9	849	20
41	625	17	375	786	26	214	160	10	840	19
42	642	16	358	812	26	188	170	10	830	18
43	658	16	342	838	26	162	180	10	820	17
44	674	17	326	864	26	136	190	9	810	16
45	691	16	309	890	26	110	199	10	801	15
46	707	16	293	916	26	084	209	10	791	14
47	723	16	277	942	26	058	219	10	781	13
48	739	17	261	968	26	032	229	10	771	12
49	756	16	244	994	26	006	239	9	761	11
50	772	16	228	89020	26	10980	248	10	752	10
51	788	17	212	046	27	954	258	10	742	9
52	805	16	195	073	26	927	268	10	732	8
53	821	16	179	099	26	901	278	10	722	7
54	837	16	163	125	26	875	288	10	712	6
55	853	16	147	151	26	849	298	9	702	5
56	869	17	131	177	26	823	307	10	693	4
57	886	16	114	203	26	797	317	10	683	3
58	902	16	098	229	26	771	327	10	673	2
59	918	16	082	255	26	745	337	10	663	1
60	78934		21066	89281		10719	10347		89653	0
′	9. l cos	d 1′	10. l sec	9. l cot	d 1′	10. l tan	10. l csc	d 1′	9. l sin	′

″	Proportional Parts 27	26	17	16	10	9
0	0	0	0	0	0	0
1	0	0	0	0	0	0
2	1	1	1	1	0	0
3	1	1	1	1	0	0
4	2	2	1	1	1	1
5	2	2	1	1	1	1
6	3	3	2	2	1	1
7	3	3	2	2	1	1
8	4	3	2	2	1	1
9	4	4	3	2	2	1
10	4	4	3	3	2	2
11	5	5	3	3	2	2
12	5	5	3	3	2	2
13	6	6	4	3	2	2
14	6	6	4	4	2	2
15	7	6	4	4	2	2
16	7	7	5	4	3	2
17	8	7	5	5	3	3
18	8	8	5	5	3	3
19	9	8	5	5	3	3
20	9	9	6	5	3	3
21	9	9	6	6	4	3
22	10	10	6	6	4	3
23	10	10	7	6	4	3
24	11	10	7	6	4	4
25	11	11	7	7	4	4
26	12	11	7	7	4	4
27	12	12	8	7	4	4
28	13	12	8	7	5	4
29	13	13	8	8	5	4
30	14	13	8	8	5	4
31	14	13	9	8	5	5
32	14	14	9	9	5	5
33	15	14	9	9	6	5
34	15	15	10	9	6	5
35	16	15	10	9	6	5
36	16	16	10	10	6	5
37	17	16	10	10	6	6
38	17	16	11	10	6	6
39	18	17	11	10	6	6
40	18	17	11	11	7	6
41	18	18	12	11	7	6
42	19	18	12	11	7	6
43	19	19	12	11	7	6
44	20	19	12	12	7	7
45	20	20	13	12	8	7
46	21	20	13	12	8	7
47	21	20	13	13	8	7
48	22	21	14	13	8	7
49	22	21	14	13	8	7
50	22	22	14	13	8	8
51	23	22	14	14	8	8
52	23	23	15	14	9	8
53	24	23	15	14	9	8
54	24	23	15	14	9	8
55	25	24	16	15	9	8
56	25	24	16	15	9	8
57	26	25	16	15	10	9
58	26	25	16	15	10	9
59	27	26	17	16	10	9
60	27	26	17	16	10	9
″	27	26	17	16	10	9

Proportional Parts

′	l sin 9.	d 1′	l csc 10.	l tan 9.	d 1′	l cot 10.	l sec 10.	d 1′	l cos 9.	′
0	78934	16	21066	89281	26	10719	10347	10	89653	60
1	950	17	050	307	26	693	357	10	643	59
2	967	16	033	333	26	667	367	9	633	58
3	983	16	017	359	26	641	376	10	624	57
4	999	16	001	385	26	615	386	10	614	56
5	79015	16	20985	411	26	589	396	10	604	55
6	031	16	969	437	26	563	406	10	594	54
7	047	16	953	463	26	537	416	10	584	53
8	063	16	937	489	26	511	426	10	574	52
9	079	16	921	515	26	485	436	10	564	51
10	095	16	905	541	26	459	446	10	554	50
11	111	17	889	567	26	433	456	10	544	49
12	128	16	872	593	26	407	466	10	534	48
13	144	16	856	619	26	381	476	10	524	47
14	160	16	840	645	26	355	486	10	514	46
15	176	16	824	671	26	329	496	9	504	45
16	192	16	808	697	26	303	505	10	495	44
17	208	16	792	723	26	277	515	10	485	43
18	224	16	776	749	26	251	525	10	475	42
19	240	16	760	775	26	225	535	10	465	41
20	256	16	744	801	26	199	545	10	455	40
21	272	16	728	827	26	173	555	10	445	39
22	288	16	712	853	26	147	565	10	435	38
23	304	15	696	879	26	121	575	10	425	37
24	319	16	681	905	26	095	585	10	415	36
25	335	16	665	931	26	069	595	10	405	35
26	351	16	649	957	26	043	605	10	395	34
27	367	16	633	983	26	017	615	10	385	33
28	383	16	617	90009	26	09991	625	11	375	32
29	399	16	601	035	26	965	636	10	364	31
30	79415	16	20585	90061	25	09939	10646	10	89354	30
31	431	16	569	086	26	914	656	10	344	29
32	447	16	553	112	26	888	666	10	334	28
33	463	15	537	138	26	862	676	10	324	27
34	478	16	522	164	26	836	686	10	314	26
35	494	16	506	190	26	810	696	10	304	25
36	510	16	490	216	26	784	706	10	294	24
37	526	16	474	242	26	758	716	10	284	23
38	542	16	458	268	26	732	726	10	274	22
39	558	15	442	294	26	706	736	10	264	21
40	573	16	427	320	26	680	746	10	254	20
41	589	16	411	346	25	654	756	11	244	19
42	605	16	395	371	26	629	767	10	233	18
43	621	15	379	397	26	603	777	10	223	17
44	636	16	364	423	26	577	787	10	213	16
45	652	16	348	449	26	551	797	10	203	15
46	668	16	332	475	26	525	807	10	193	14
47	684	15	316	501	26	499	817	10	183	13
48	699	16	301	527	26	473	827	11	173	12
49	715	16	285	553	25	447	838	10	162	11
50	731	15	269	578	26	422	848	10	152	10
51	746	16	254	604	26	396	858	10	142	9
52	762	16	238	630	26	370	868	10	132	8
53	778	15	222	656	26	344	878	10	122	7
54	793	16	207	682	26	318	888	11	112	6
55	809	16	191	708	26	292	899	10	101	5
56	825	15	175	734	25	266	909	10	091	4
57	840	16	160	759	26	241	919	10	081	3
58	856	16	144	785	26	215	929	11	071	2
59	872	15	128	811	26	189	940	10	060	1
60	79887		20113	90837		09163	10950		89050	0
′	9. l cos	d 1′	10. l sec	9. l cot	d 1′	10. l tan	10. l csc	d 1′	9. l sin	′

Proportional Parts

″	26	25	17	16	15	11	10	9
0	0	0	0	0	0	0	0	0
1	0	0	0	0	0	0	0	0
2	1	1	1	1	0	0	0	0
3	1	1	1	1	1	1	0	0
4	2	2	1	1	1	1	1	1
5	2	2	1	1	1	1	1	1
6	3	2	2	2	2	1	1	1
7	3	3	2	2	2	1	1	1
8	3	3	2	2	2	1	1	1
9	4	4	3	2	2	2	2	1
10	4	4	3	3	2	2	2	2
11	5	5	3	3	3	2	2	2
12	5	5	3	3	3	2	2	2
13	6	5	4	3	3	2	2	2
14	6	6	4	4	4	3	2	2
15	6	6	4	4	4	3	2	2
16	7	7	5	4	4	3	3	2
17	7	7	5	5	4	3	3	3
18	8	8	5	5	4	3	3	3
19	8	8	5	5	5	3	3	3
20	9	8	6	5	5	4	3	3
21	9	9	6	6	5	4	4	3
22	10	9	6	6	6	4	4	3
23	10	10	7	6	6	4	4	3
24	10	10	7	6	6	4	4	4
25	11	10	7	7	6	5	4	4
26	11	11	7	7	6	5	4	4
27	12	11	8	7	7	5	4	4
28	12	12	8	7	7	5	5	4
29	13	12	8	8	7	5	5	4
30	13	12	8	8	8	6	5	4
31	13	13	9	8	8	6	5	5
32	14	13	9	9	8	6	5	5
33	14	14	9	9	8	6	6	5
34	15	14	10	9	8	6	6	5
35	15	15	10	9	9	6	6	5
36	16	15	10	10	9	7	6	5
37	16	15	10	10	9	7	6	6
38	16	16	11	10	10	7	6	6
39	17	16	11	10	10	7	6	6
40	17	17	11	11	10	7	7	6
41	18	17	12	11	10	8	7	6
42	18	18	12	11	10	8	7	6
43	19	18	12	11	11	8	7	6
44	19	18	12	12	11	8	7	7
45	20	19	13	12	11	8	8	7
46	20	19	13	12	12	8	8	7
47	20	20	13	13	12	9	8	7
48	21	20	14	13	12	9	8	7
49	21	20	14	13	12	9	8	7
50	22	21	14	13	12	9	8	8
51	22	21	14	14	13	9	8	8
52	23	22	15	14	13	10	9	8
53	23	22	15	14	13	10	9	8
54	23	22	15	14	14	10	9	8
55	24	23	16	15	14	10	9	8
56	24	23	16	15	14	10	9	8
57	25	24	16	15	14	10	10	9
58	25	24	16	15	14	11	10	9
59	26	25	17	16	15	11	10	9
60	26	25	17	16	15	11	10	9
″	26	25	17	16	15	11	10	9

Proportional Parts

128° 51°

′	l sin 9.	d 1′	l csc 10.	l tan 9.	d 1′	l cot 10.	l sec 10.	d 1′	l cos 9.	′
0	79887	16	20113	90837	26	09163	10950	10	89050	60
1	903	15	097	863	26	137	960	10	040	59
2	918	16	082	889	25	111	970	10	030	58
3	934	16	066	914	26	086	980	11	020	57
4	950	15	050	940	26	060	991	10	009	56
5	965	16	035	966	26	034	11001	10	88999	55
6	981	15	019	992	26	008	011	11	989	54
7	996	16	004	91018	25	08982	022	10	978	53
8	80012	15	19988	043	26	957	032	10	968	52
9	027	16	973	069	26	931	042	10	958	51
10	043	15	957	095	26	905	052	11	948	50
11	058	16	942	121	26	879	063	10	937	49
12	074	15	926	147	25	853	073	10	927	48
13	089	16	911	172	26	828	083	11	917	47
14	105	15	895	198	26	802	094	10	906	46
15	120	16	880	224	26	776	104	10	896	45
16	136	15	864	250	26	750	114	11	886	44
17	151	15	849	276	25	724	125	10	875	43
18	166	16	834	301	26	699	135	10	865	42
19	182	15	818	327	26	673	145	11	855	41
20	197	16	803	353	26	647	156	10	844	40
21	213	15	787	379	25	621	166	10	834	39
22	228	16	772	404	26	596	176	11	824	38
23	244	15	756	430	26	570	187	10	813	37
24	259	15	741	456	26	544	197	10	803	36
25	274	16	726	482	25	518	207	11	793	35
26	290	15	710	507	26	493	218	10	782	34
27	305	15	695	533	26	467	228	11	772	33
28	320	16	680	559	26	441	239	10	761	32
29	336	15	664	585	25	415	249	10	751	31
30	80351	15	19649	91610	26	08390	11259	11	88741	30
31	366	16	634	636	26	364	270	10	730	29
32	382	15	618	662	26	338	280	11	720	28
33	397	15	603	688	25	312	291	10	709	27
34	412	16	588	713	26	287	301	11	699	26
35	428	15	572	739	26	261	312	10	688	25
36	443	15	557	765	26	235	322	10	678	24
37	458	15	542	791	25	209	332	11	668	23
38	473	16	527	816	26	184	343	10	657	22
39	489	15	511	842	26	158	353	11	647	21
40	504	15	496	868	25	132	364	10	636	20
41	519	15	481	893	26	107	374	11	626	19
42	534	16	466	919	26	081	385	10	615	18
43	550	15	450	945	26	055	395	11	605	17
44	565	15	435	971	25	029	406	10	594	16
45	580	15	420	996	26	004	416	11	584	15
46	595	15	405	92022	26	07978	427	10	573	14
47	610	15	390	048	25	952	437	11	563	13
48	625	16	375	073	26	927	448	10	552	12
49	641	15	359	099	26	901	458	11	542	11
50	656	15	344	125	25	875	469	10	531	10
51	671	15	329	150	26	850	479	11	521	9
52	686	15	314	176	26	824	490	11	510	8
53	701	15	299	202	25	798	501	10	499	7
54	716	15	284	227	26	773	511	11	489	6
55	731	15	269	253	26	747	522	10	478	5
56	746	16	254	279	25	721	532	11	468	4
57	762	15	238	304	26	696	543	10	457	3
58	777	15	223	330	26	670	553	11	447	2
59	792	15	208	356	25	644	564	11	436	1
60	80807		19193	92381		07619	11575		88425	0
′	9. l cos	d 1′	10. l sec	9. l cot	d 1′	10. l tan	10. l csc	d 1′	9. l sin	′

″	Proportional Parts 26	25	16	15	11	10
0	0	0	0	0	0	0
1	0	0	0	0	0	0
2	1	1	1	0	0	0
3	1	1	1	1	1	0
4	2	2	1	1	1	1
5	2	2	1	1	1	1
6	3	2	2	2	1	1
7	3	3	2	2	1	1
8	3	3	2	2	1	1
9	4	4	2	2	2	2
10	4	4	3	2	2	2
11	5	5	3	3	2	2
12	5	5	3	3	2	2
13	6	5	3	3	2	2
14	6	6	4	4	3	2
15	6	6	4	4	3	2
16	7	7	4	4	3	3
17	7	7	5	4	3	3
18	8	8	5	4	3	3
19	8	8	5	5	3	3
20	9	8	5	5	4	3
21	9	9	6	5	4	4
22	10	9	6	6	4	4
23	10	10	6	6	4	4
24	10	10	6	6	4	4
25	11	10	7	6	5	4
26	11	11	7	6	5	4
27	12	11	7	7	5	4
28	12	12	7	7	5	5
29	13	12	8	7	5	5
30	13	12	8	8	6	5
31	13	13	8	8	6	5
32	14	13	9	8	6	5
33	14	14	9	8	6	6
34	15	14	9	8	6	6
35	15	15	9	9	6	6
36	16	15	10	9	7	6
37	16	15	10	9	7	6
38	16	16	10	10	7	6
39	17	16	10	10	7	6
40	17	17	11	10	7	7
41	18	17	11	10	8	7
42	18	18	11	10	8	7
43	19	18	11	11	8	7
44	19	18	12	11	8	7
45	20	19	12	11	8	8
46	20	19	12	12	8	8
47	20	20	13	12	9	8
48	21	20	13	12	9	8
49	21	20	13	12	9	8
50	22	21	13	12	9	8
51	22	21	14	13	9	8
52	23	22	14	13	10	9
53	23	22	14	13	10	9
54	23	22	14	14	10	9
55	24	23	15	14	10	9
56	24	23	15	14	10	9
57	25	24	15	14	10	10
58	25	24	15	14	11	10
59	26	25	16	15	11	10
60	26	25	16	15	11	10
″	26 Proportional Parts	25	16	15	11	10

′	l sin 9.	d 1′	l csc 10.	l tan 9.	d 1′	l cot 10.	l sec 10.	d 1′	l cos 9.	′
0	**80807**	15	**19193**	**92381**	26	**07619**	**11575**	10	**88425**	60
1	822	15	178	407	26	593	585	11	415	59
2	837	15	163	433	25	567	596	10	404	58
3	852	15	148	458	26	542	606	11	394	57
4	867	15	133	484	26	516	617	11	383	56
5	882	15	118	510	25	490	628	10	372	55
6	897	15	103	535	26	465	638	11	362	54
7	912	15	088	561	26	439	649	11	351	53
8	927	15	073	587	25	413	660	10	340	52
9	942	15	058	612	26	388	670	11	330	51
10	957	15	043	638	25	362	681	11	319	50
11	972	15	028	663	26	337	692	10	308	49
12	987	15	013	689	26	311	702	11	298	48
13	**81002**	15	**18998**	715	25	285	713	11	287	47
14	017	15	983	740	26	260	724	10	276	46
15	032	15	968	766	26	234	734	11	266	45
16	047	14	953	792	25	208	745	11	255	44
17	061	15	939	817	26	183	756	10	244	43
18	076	15	924	843	25	157	766	11	234	42
19	091	15	909	868	26	132	777	11	223	41
20	106	15	894	894	26	106	788	11	212	40
21	121	15	879	920	25	080	799	10	201	39
22	136	15	864	945	26	055	809	11	191	38
23	151	15	849	971	25	029	820	11	180	37
24	166	14	834	996	26	004	831	11	169	36
25	180	15	820	**93022**	26	**06978**	842	10	158	35
26	195	15	805	048	25	952	852	11	148	34
27	210	15	790	073	26	927	863	11	137	33
28	225	15	775	099	25	901	874	11	126	32
29	240	14	760	124	26	876	885	10	115	31
30	**81254**	15	**18746**	**93150**	25	**06850**	**11895**	11	**88105**	30
31	269	15	731	175	26	825	906	11	094	29
32	284	15	716	201	26	799	917	11	083	28
33	299	15	701	227	25	773	928	11	072	27
34	314	14	686	252	26	748	939	10	061	26
35	328	15	672	278	25	722	949	11	051	25
36	343	15	657	303	26	697	960	11	040	24
37	358	14	642	329	25	671	971	11	029	23
38	372	15	628	354	26	646	982	11	018	22
39	387	15	613	380	26	620	993	11	007	21
40	402	15	598	406	25	594	**12004**	11	**87996**	20
41	417	14	583	431	26	569	015	10	985	19
42	431	15	569	457	25	543	025	11	975	18
43	446	15	554	482	26	518	036	11	964	17
44	461	14	539	508	25	492	047	11	953	16
45	475	15	525	533	26	467	058	11	942	15
46	490	15	510	559	25	441	069	11	931	14
47	505	14	495	584	26	416	080	11	920	13
48	519	15	481	610	26	390	091	11	909	12
49	534	15	466	636	25	364	102	11	898	11
50	549	14	451	661	26	339	113	10	887	10
51	563	15	437	687	25	313	123	11	877	9
52	578	14	422	712	26	288	134	11	866	8
53	592	15	408	738	25	262	145	11	855	7
54	607	15	393	763	26	237	156	11	844	6
55	622	14	378	789	25	211	167	11	833	5
56	636	15	364	814	26	186	178	11	822	4
57	651	14	349	840	25	160	189	11	811	3
58	665	15	335	865	26	135	200	11	800	2
59	680	14	320	891	25	109	211	11	789	1
60	**81694**		**18306**	**93916**		**06084**	**12222**		**87778**	0
′	9. l cos	d 1′	10. l sec	9. l cot	d 1′	10. l tan	10. l csc	d 1′	9. l sin	′

Proportional Parts

″	26	25	15	14	11	10
0	0	0	0	0	0	0
1	0	0	0	0	0	0
2	1	1	0	0	0	0
3	1	1	1	1	1	0
4	2	2	1	1	1	1
5	2	2	1	1	1	1
6	3	2	2	1	1	1
7	3	3	2	2	1	1
8	3	3	2	2	1	1
9	4	4	2	2	2	2
10	4	4	2	2	2	2
11	5	5	3	3	2	2
12	5	5	3	3	2	2
13	6	5	3	3	2	2
14	6	6	4	3	3	2
15	6	6	4	4	3	2
16	7	7	4	4	3	3
17	7	7	4	4	3	3
18	8	8	4	4	3	3
19	8	8	5	4	3	3
20	9	8	5	5	4	3
21	9	9	5	5	4	4
22	10	9	6	5	4	4
23	10	10	6	5	4	4
24	10	10	6	6	4	4
25	11	10	6	6	5	4
26	11	11	6	6	5	4
27	12	11	7	6	5	4
28	12	12	7	7	5	5
29	13	12	7	7	5	5
30	13	12	8	7	6	5
31	13	13	8	7	6	5
32	14	13	8	7	6	5
33	14	14	8	8	6	6
34	15	14	8	8	6	6
35	15	15	9	8	6	6
36	16	15	9	8	7	6
37	16	15	9	9	7	6
38	16	16	10	9	7	6
39	17	16	10	9	7	6
40	17	17	10	9	7	7
41	18	17	10	10	8	7
42	18	18	10	10	8	7
43	19	18	11	10	8	7
44	19	18	11	10	8	7
45	20	19	11	10	8	8
46	20	19	12	11	8	8
47	20	20	12	11	9	8
48	21	20	12	11	9	8
49	21	20	12	11	9	8
50	22	21	12	12	9	8
51	22	21	13	12	9	8
52	23	22	13	12	10	9
53	23	22	13	12	10	9
54	23	22	14	13	10	9
55	24	23	14	13	10	9
56	24	23	14	13	10	9
57	25	24	14	13	10	10
58	25	24	14	14	11	10
59	26	25	15	14	11	10
60	26	25	15	14	11	10
″	26	25	15	14	11	10

Proportional Parts

′	l sin 9.	d 1′	l csc 10.	l tan 9.	d 1′	l cot 10.	l sec 10.	d 1′	l cos 9.	′
0	81694	15	18306	93916	26	06084	12222	11	87778	60
1	709	14	291	942	25	058	233	11	767	59
2	723	15	277	967	26	033	244	11	756	58
3	738	14	262	993	25	007	255	11	745	57
4	752	15	248	94018	26	05982	266	11	734	56
5	767	14	233	044	25	956	277	11	723	55
6	781	15	219	069	26	931	288	11	712	54
7	796	14	204	095	25	905	299	11	701	53
8	810	15	190	120	26	880	310	11	690	52
9	825	14	175	146	25	854	321	11	679	51
10	839	15	161	171	26	829	332	11	668	50
11	854	14	146	197	25	803	343	11	657	49
12	868	14	132	222	26	778	354	11	646	48
13	882	15	118	248	25	752	365	11	635	47
14	897	14	103	273	26	727	376	11	624	46
15	911	15	089	299	25	701	387	12	613	45
16	926	14	074	324	26	676	399	11	601	44
17	940	15	060	350	25	650	410	11	590	43
18	955	14	045	375	26	625	421	11	579	42
19	969	14	031	401	25	599	432	11	568	41
20	983	15	017	426	26	574	443	11	557	40
21	998	14	002	452	25	548	454	11	546	39
22	82012	14	17988	477	26	523	465	11	535	38
23	026	15	974	503	25	497	476	11	524	37
24	041	14	959	528	26	472	487	12	513	36
25	055	14	945	554	25	446	499	11	501	35
26	069	15	931	579	25	421	510	11	490	34
27	084	14	916	604	26	396	521	11	479	33
28	098	14	902	630	25	370	532	11	468	32
29	112	14	888	655	26	345	543	11	457	31
30	82126	15	17874	94681	25	05319	12554	12	87446	30
31	141	14	859	706	26	294	566	11	434	29
32	155	14	845	732	25	268	577	11	423	28
33	169	15	831	757	26	243	588	11	412	27
34	184	14	816	783	25	217	599	11	401	26
35	198	14	802	808	26	192	610	12	390	25
36	212	14	788	834	25	166	622	11	378	24
37	226	14	774	859	25	141	633	11	367	23
38	240	15	760	884	26	116	644	11	356	22
39	255	14	745	910	25	090	655	11	345	21
40	269	14	731	935	26	065	666	12	334	20
41	283	14	717	961	25	039	678	11	322	19
42	297	14	703	986	26	014	689	11	311	18
43	311	15	689	95012	25	04988	700	12	300	17
44	326	14	674	037	25	963	712	11	288	16
45	340	14	660	062	26	938	723	11	277	15
46	354	14	646	088	25	912	734	11	266	14
47	368	14	632	113	26	887	745	12	255	13
48	382	14	618	139	25	861	757	11	243	12
49	396	14	604	164	26	836	768	11	232	11
50	410	14	590	190	25	810	779	12	221	10
51	424	15	576	215	25	785	791	11	209	9
52	439	14	561	240	26	760	802	11	198	8
53	453	14	547	266	25	734	813	12	187	7
54	467	14	533	291	26	709	825	11	175	6
55	481	14	519	317	25	683	836	11	164	5
56	495	14	505	342	26	658	847	12	153	4
57	509	14	491	368	25	632	859	11	141	3
58	523	14	477	393	25	607	870	11	130	2
59	537	14	463	418	26	582	881	12	119	1
60	82551		17449	95444		04556	12893		87107	0
′	9. l cos	d 1′	10. l sec	9. l cot	d 1′	10. l tan	10. l csc	d 1′	9. l sin	′

″	26	25	15	14	12	11
	Proportional Parts					
0	0	0	0	0	0	0
1	0	0	0	0	0	0
2	1	1	0	0	0	0
3	1	1	1	1	1	1
4	2	2	1	1	1	1
5	2	2	1	1	1	1
6	3	2	2	1	1	1
7	3	3	2	2	1	1
8	3	3	2	2	2	1
9	4	4	2	2	2	2
10	4	4	2	2	2	2
11	5	5	3	3	2	2
12	5	5	3	3	2	2
13	6	5	3	3	3	2
14	6	6	4	3	3	3
15	6	6	4	4	3	3
16	7	7	4	4	3	3
17	7	7	4	4	3	3
18	8	8	4	4	4	3
19	8	8	5	4	4	3
20	9	8	5	5	4	4
21	9	9	5	5	4	4
22	10	9	6	5	4	4
23	10	10	6	5	5	4
24	10	10	6	6	5	4
25	11	10	6	6	5	5
26	11	11	6	6	5	5
27	12	11	7	6	5	5
28	12	12	7	7	6	5
29	13	12	7	7	6	5
30	13	12	8	7	6	6
31	13	13	8	7	6	6
32	14	13	8	7	6	6
33	14	14	8	8	7	6
34	15	14	8	8	7	6
35	15	15	9	8	7	6
36	16	15	9	8	7	7
37	16	15	9	9	7	7
38	16	16	10	9	8	7
39	17	16	10	9	8	7
40	17	17	10	9	8	7
41	18	17	10	10	8	8
42	18	18	10	10	8	8
43	19	18	11	10	9	8
44	19	18	11	10	9	8
45	20	19	11	10	9	8
46	20	19	12	11	9	8
47	20	20	12	11	9	9
48	21	20	12	11	10	9
49	21	20	12	11	10	9
50	22	21	12	12	10	9
51	22	21	13	12	10	9
52	23	22	13	12	10	10
53	23	22	13	12	11	10
54	23	22	14	13	11	10
55	24	23	14	13	11	10
56	24	23	14	13	11	10
57	25	24	14	13	11	10
58	25	24	14	14	12	11
59	26	25	15	14	12	11
60	26	25	15	14	12	11
″	26	25	15	14	12	11
	Proportional Parts					

'	*l* sin 9.	d 1'	*l* csc 10.	*l* tan 9.	d 1'	*l* cot 10.	*l* sec 10.	d 1'	*l* cos 9.	'
0	82551	14	17449	95444	25	04556	12893	11	87107	60
1	565	14	435	469	26	531	904	11	096	59
2	579	14	421	495	25	505	915	12	085	58
3	593	14	407	520	25	480	927	11	073	57
4	607	14	393	545	26	455	938	12	062	56
5	621	14	379	571	25	429	950	11	050	55
6	635	14	365	596	26	404	961	11	039	54
7	649	14	351	622	25	378	972	12	028	53
8	663	14	337	647	25	353	984	11	016	52
9	677	14	323	672	26	328	995	12	005	51
0	691	14	309	698	25	302	13007	11	86993	50
1	705	14	295	723	25	277	018	12	982	49
2	719	14	281	748	26	252	030	11	970	48
3	733	14	267	774	25	226	041	12	959	47
4	747	14	253	799	26	201	053	11	947	46
5	761	14	239	825	25	175	064	12	936	45
6	775	13	225	850	25	150	076	11	924	44
7	788	14	212	875	26	125	087	11	913	43
8	802	14	198	901	25	099	098	12	902	42
9	816	14	184	926	26	074	110	11	890	41
0	830	14	170	952	25	048	121	12	879	40
1	844	14	156	977	25	023	133	12	867	39
2	858	14	142	96002	26	03998	145	11	855	38
3	872	13	128	028	25	972	156	12	844	37
4	885	14	115	053	25	947	168	11	832	36
5	899	14	101	078	26	922	179	12	821	35
6	913	14	087	104	25	896	191	11	809	34
7	927	14	073	129	26	871	202	12	798	33
8	941	14	059	155	25	845	214	11	786	32
9	955	13	045	180	25	820	225	12	775	31
0	82968	14	17032	96205	26	03795	13237	11	86763	30
1	982	14	018	231	25	769	248	12	752	29
2	996	14	004	256	25	744	260	12	740	28
3	83010	13	16990	281	26	719	272	11	728	27
4	023	14	977	307	25	693	283	12	717	26
5	037	14	963	332	25	668	295	11	705	25
6	051	14	949	357	26	643	306	12	694	24
7	065	13	935	383	25	617	318	12	682	23
8	078	14	922	408	25	592	330	11	670	22
9	092	14	908	433	26	567	341	12	659	21
0	106	14	894	459	25	541	353	12	647	20
1	120	13	880	484	26	516	365	11	635	19
2	133	14	867	510	25	490	376	12	624	18
3	147	14	853	535	25	465	388	12	612	17
4	161	13	839	560	26	440	400	11	600	16
5	174	14	826	586	25	414	411	12	589	15
6	188	14	812	611	25	389	423	12	577	14
7	202	13	798	636	26	364	435	11	565	13
8	215	14	785	662	25	338	446	12	554	12
9	229	13	771	687	25	313	458	12	542	11
0	242	14	758	712	26	288	470	12	530	10
1	256	14	744	738	25	262	482	11	518	9
2	270	13	730	763	25	237	493	12	507	8
3	283	14	717	788	26	212	505	12	495	7
4	297	13	703	814	25	186	517	11	483	6
5	310	14	690	839	25	161	528	12	472	5
6	324	14	676	864	26	136	540	12	460	4
7	338	13	662	890	25	110	552	12	448	3
8	351	14	649	915	25	085	564	11	436	2
9	365	13	635	940	26	060	575	12	425	1
0	83378		16622	96966		03034	13587		86413	0
	9. *l* cos	d 1'	10. *l* sec	9. *l* cot	d 1'	10. *l* tan	10. *l* csc	d 1'	9. *l* sin	'

"	Proportional Parts 26	25	14	13	12	11
0	0	0	0	0	0	0
1	0	0	0	0	0	0
2	1	1	0	0	0	0
3	1	1	1	1	1	1
4	2	2	1	1	1	1
5	2	2	1	1	1	1
6	3	2	1	1	1	1
7	3	3	2	2	1	1
8	3	3	2	2	2	1
9	4	4	2	2	2	2
10	4	4	2	2	2	2
11	5	5	3	2	2	2
12	5	5	3	3	2	2
13	6	5	3	3	3	2
14	6	6	3	3	3	3
15	6	6	4	3	3	3
16	7	7	4	3	3	3
17	7	7	4	4	3	3
18	8	8	4	4	4	3
19	8	8	4	4	4	3
20	9	8	5	4	4	4
21	9	9	5	5	4	4
22	10	9	5	5	4	4
23	10	10	5	5	5	4
24	10	10	6	5	5	4
25	11	10	6	5	5	5
26	11	11	6	6	5	5
27	12	11	6	6	5	5
28	12	12	7	6	6	5
29	13	12	7	6	6	5
30	13	12	7	6	6	6
31	13	13	7	7	6	6
32	14	13	7	7	6	6
33	14	14	8	7	7	6
34	15	14	8	7	7	6
35	15	15	8	8	7	6
36	16	15	8	8	7	7
37	16	15	9	8	7	7
38	16	16	9	8	8	7
39	17	16	9	8	8	7
40	17	17	9	9	8	7
41	18	17	10	9	8	8
42	18	18	10	9	8	8
43	19	18	10	9	9	8
44	19	18	10	10	9	8
45	20	19	10	10	9	8
46	20	19	11	10	9	8
47	20	20	11	10	9	9
48	21	20	11	10	10	9
49	21	20	11	11	10	9
50	22	21	12	11	10	9
51	22	21	12	11	10	9
52	23	22	12	11	10	10
53	23	22	12	11	11	10
54	23	22	13	12	11	10
55	24	23	13	12	11	10
56	24	23	13	12	11	10
57	25	24	13	12	11	10
58	25	24	14	13	12	11
59	26	25	14	13	12	11
60	26	25	14	13	12	11
"	26	25	14	13	12	11

Proportional Parts

′	l sin 9.	d 1′	l csc 10.	l tan 9.	d 1′	l cot 10.	l sec 10.	d 1′	l cos 9.	′
0	**83**378	14	**16**622	**96**966	25	**03**034	**13**587	12	**86**413	60
1	392	13	608	991	25	009	599	12	401	59
2	405	14	595	**97**016	26	**02**984	611	12	389	58
3	419	13	581	042	25	958	623	11	377	57
4	432	14	568	067	25	933	634	12	366	56
5	446	13	554	092	26	908	646	12	354	55
6	459	14	541	118	25	882	658	12	342	54
7	473	13	527	143	25	857	670	12	330	53
8	486	14	514	168	25	832	682	12	318	52
9	500	13	500	193	26	807	694	11	306	51
10	513	14	487	219	25	781	705	12	295	50
11	527	13	473	244	25	756	717	12	283	49
12	540	14	460	269	26	731	729	12	271	48
13	554	13	446	295	25	705	741	12	259	47
14	567	14	433	320	25	680	753	12	247	46
15	581	13	419	345	26	655	765	12	235	45
16	594	14	406	371	25	629	777	12	223	44
17	608	13	392	396	25	604	789	11	211	43
18	621	13	379	421	26	579	800	12	200	42
19	634	14	366	447	25	553	812	12	188	41
20	648	13	352	472	25	528	824	12	176	40
21	661	13	339	497	26	503	836	12	164	39
22	674	14	326	523	25	477	848	12	152	38
23	688	13	312	548	25	452	860	12	140	37
24	701	14	299	573	25	427	872	12	128	36
25	715	13	285	598	26	402	884	12	116	35
26	728	13	272	624	25	376	896	12	104	34
27	741	14	259	649	25	351	908	12	092	33
28	755	13	245	674	26	326	920	12	080	32
29	768	13	232	700	25	300	932	12	068	31
30	**83**781	14	**16**219	**97**725	25	**02**275	**13**944	12	**86**056	30
31	795	13	205	750	26	250	956	12	044	29
32	808	13	192	776	25	224	968	12	032	28
33	821	13	179	801	25	199	980	12	020	27
34	834	14	166	826	25	174	992	12	008	26
35	848	13	152	851	26	149	**14**004	12	**85**996	25
36	861	13	139	877	25	123	016	12	984	24
37	874	13	126	902	25	098	028	12	972	23
38	887	14	113	927	26	073	040	12	960	22
39	901	13	099	953	25	047	052	12	948	21
40	914	13	086	978	25	022	064	12	936	20
41	927	13	073	**98**003	26	**01**997	076	12	924	19
42	940	14	060	029	25	971	088	12	912	18
43	954	13	046	054	25	946	100	12	900	17
44	967	13	033	079	25	921	112	12	888	16
45	980	13	020	104	26	896	124	12	876	15
46	993	13	007	130	25	870	136	13	864	14
47	**84**006	14	**15**994	155	25	845	149	12	851	13
48	020	13	980	180	26	820	161	12	839	12
49	033	13	967	206	25	794	173	12	827	11
50	046	13	954	231	25	769	185	12	815	10
51	059	13	941	256	25	744	197	12	803	9
52	072	13	928	281	26	719	209	12	791	8
53	085	13	915	307	25	693	221	13	779	7
54	098	14	902	332	25	668	234	12	766	6
55	112	13	888	357	26	643	246	12	754	5
56	125	13	875	383	25	617	258	12	742	4
57	138	13	862	408	25	592	270	12	730	3
58	151	13	849	433	25	567	282	12	718	2
59	164	13	836	458	26	542	294	13	706	1
60	**84**177		**15**823	**98**484		**01**516	**14**307		**85**693	0
′	9. l cos	d 1′	10. l sec	9. l cot	d 1′	10. l tan	10. l csc	d 1′	9. l sin	′

133° 46°

″	26	25	14	13	12	11
	Proportional Parts					
0	0	0	0	0	0	0
1	0	0	0	0	0	0
2	1	1	0	0	0	0
3	1	1	1	1	1	1
4	2	2	1	1	1	1
5	2	2	1	1	1	1
6	3	2	1	1	1	1
7	3	3	2	2	1	1
8	3	3	2	2	2	1
9	4	4	2	2	2	2
10	4	4	2	2	2	2
11	5	5	3	2	2	2
12	5	5	3	3	2	2
13	6	5	3	3	3	2
14	6	6	3	3	3	3
15	6	6	4	3	3	3
16	7	7	4	3	3	3
17	7	7	4	4	3	3
18	8	8	4	4	4	3
19	8	8	4	4	4	3
20	9	8	5	4	4	4
21	9	9	5	5	4	4
22	10	9	5	5	4	4
23	10	10	5	5	5	4
24	10	10	6	5	5	4
25	11	10	6	5	5	5
26	11	11	6	6	5	5
27	12	11	6	6	5	5
28	12	12	7	6	6	5
29	13	12	7	6	6	5
30	13	12	7	6	6	6
31	13	13	7	7	6	6
32	14	13	7	7	6	6
33	14	14	8	7	7	6
34	15	14	8	7	7	6
35	15	15	8	8	7	6
36	16	15	8	8	7	7
37	16	15	9	8	7	7
38	16	16	9	8	8	7
39	17	16	9	8	8	7
40	17	17	9	9	8	7
41	18	17	10	9	8	8
42	18	18	10	9	8	8
43	19	18	10	9	9	8
44	19	18	10	10	9	8
45	20	19	10	10	9	8
46	20	19	11	10	9	8
47	20	20	11	10	9	9
48	21	20	11	10	10	9
49	21	20	11	11	10	9
50	22	21	12	11	10	9
51	22	21	12	11	10	9
52	23	22	12	11	10	10
53	23	22	12	11	11	10
54	23	22	13	12	11	10
55	24	23	13	12	11	10
56	24	23	13	12	11	10
57	25	24	13	12	11	10
58	25	24	14	13	12	11
59	26	25	14	13	12	11
60	26	25	14	13	12	11
″	26	25	14	13	12	11
	Proportional Parts					

′	*l* sin 9.	d 1′	*l* csc 10.	*l* tan 9.	d 1′	*l* cot 10.	*l* sec 10.	d 1′	*l* cos 9.	′
0	84177	13	15823	98484	25	01516	14307	12	85693	60
1	190	13	810	509	25	491	319	12	681	59
2	203	13	797	534	26	466	331	12	669	58
3	216	13	784	560	25	440	343	12	657	57
4	229	13	771	585	25	415	355	13	645	56
5	242	13	758	610	25	390	368	12	632	55
6	255	14	745	635	26	365	380	12	620	54
7	269	13	731	661	25	339	392	12	608	53
8	282	13	718	686	25	314	404	13	596	52
9	295	13	705	711	26	289	417	12	583	51
10	308	13	692	737	25	263	429	12	571	50
11	321	13	679	762	25	238	441	12	559	49
12	334	13	666	787	25	213	453	13	547	48
13	347	13	653	812	26	188	466	12	534	47
14	360	13	640	838	25	162	478	12	522	46
15	373	12	627	863	25	137	490	13	510	45
16	385	13	615	888	25	112	503	12	497	44
17	398	13	602	913	26	087	515	12	485	43
18	411	13	589	939	25	061	527	13	473	42
19	424	13	576	964	25	036	540	12	460	41
20	437	13	563	989	26	011	552	12	448	40
21	450	13	550	99015	25	00985	564	13	436	39
22	463	13	537	040	25	960	577	12	423	38
23	476	13	524	065	25	935	589	12	411	37
24	489	13	511	090	26	910	601	13	399	36
25	502	13	498	116	25	884	614	12	386	35
26	515	13	485	141	25	859	626	13	374	34
27	528	12	472	166	25	834	639	12	361	33
28	540	13	460	191	26	809	651	12	349	32
29	553	13	447	217	25	783	663	13	337	31
30	84566	13	15434	99242	25	00758	14676	12	85324	30
31	579	13	421	267	26	733	688	13	312	29
32	592	13	408	293	25	707	701	12	299	28
33	605	13	395	318	25	682	713	13	287	27
34	618	12	382	343	25	657	726	12	274	26
35	630	13	370	368	26	632	738	12	262	25
36	643	13	357	394	25	606	750	13	250	24
37	656	13	344	419	25	581	763	12	237	23
38	669	13	331	444	25	556	775	13	225	22
39	682	12	318	469	26	531	788	12	212	21
40	694	13	306	495	25	505	800	13	200	20
41	707	13	293	520	25	480	813	12	187	19
42	720	13	280	545	25	455	825	13	175	18
43	733	12	267	570	26	430	838	12	162	17
44	745	13	255	596	25	404	850	13	150	16
45	758	13	242	621	25	379	863	12	137	15
46	771	13	229	646	26	354	875	13	125	14
47	784	12	216	672	25	328	888	12	112	13
48	796	13	204	697	25	303	900	13	100	12
49	809	13	191	722	25	278	913	13	087	11
50	822	13	178	747	26	253	926	12	074	10
51	835	12	165	773	25	227	938	13	062	9
52	847	13	153	798	25	202	951	12	049	8
53	860	13	140	823	25	177	963	13	037	7
54	873	12	127	848	26	152	976	12	024	6
55	885	13	115	874	25	126	988	13	012	5
56	898	13	102	899	25	101	15001	13	84999	4
57	911	12	089	924	25	076	014	12	986	3
58	923	13	077	949	26	051	026	13	974	2
59	936	13	064	975	25	025	039	12	961	1
60	84949		15051	00000		00000	15051		84949	0
′	9. *l* cos	d 1′	10. *l* sec	10. *l* cot	d 1′	10. *l* tan	10. *l* csc	d 1′	9. *l* sin	′

″	Proportional Parts 26	25	14	13	12
0	0	0	0	0	0
1	0	0	0	0	0
2	1	1	0	0	0
3	1	1	1	1	1
4	2	2	1	1	1
5	2	2	1	1	1
6	3	2	1	1	1
7	3	3	2	2	1
8	3	3	2	2	2
9	4	4	2	2	2
10	4	4	2	2	2
11	5	5	3	2	2
12	5	5	3	3	2
13	6	5	3	3	3
14	6	6	3	3	3
15	6	6	4	3	3
16	7	7	4	3	3
17	7	7	4	4	3
18	8	8	4	4	4
19	8	8	4	4	4
20	9	8	5	4	4
21	9	9	5	5	4
22	10	9	5	5	4
23	10	10	5	5	5
24	10	10	6	5	5
25	11	10	6	5	5
26	11	11	6	6	5
27	12	11	6	6	5
28	12	12	7	6	6
29	13	12	7	6	6
30	13	12	7	6	6
31	13	13	7	7	6
32	14	13	7	7	6
33	14	14	8	7	7
34	15	14	8	7	7
35	15	15	8	8	7
36	16	15	8	8	7
37	16	15	9	8	7
38	16	16	9	8	8
39	17	16	9	8	8
40	17	17	9	9	8
41	18	17	10	9	8
42	18	18	10	9	8
43	19	18	10	9	9
44	19	18	10	10	9
45	20	19	10	10	9
46	20	19	11	10	9
47	20	20	11	10	9
48	21	20	11	10	10
49	21	20	11	11	10
50	22	21	12	11	10
51	22	21	12	11	10
52	23	22	12	11	10
53	23	22	12	11	11
54	23	22	13	12	11
55	24	23	13	12	11
56	24	23	13	12	11
57	25	24	13	12	11
58	25	24	14	13	12
59	26	25	14	13	12
60	26	25	14	13	12
″	26 Proportional Parts	25	14	13	12

TABLE III

NATURAL TRIGONOMETRIC FUNCTIONS

Of angles for each minute from 0° to 90°, correct
to five significant figures

0°

′	sin	tan	cot	cos	
0	.00000	.00000	∞	1.0000	**60**
1	029	029	3437.7	000	59
2	058	058	1718.9	000	58
3	087	087	1145.9	000	57
4	116	116	859.44	000	56
5	.00145	.00145	687.55̅	1.0000	**55**
6	175̅	175̅	572.96	000	54
7	204	204	491.11	000	53
8	233	233	429.72	000	52
9	262	262	381.97	000	51
10	.00291	.00291	343.77	1.0000	**50**
11	320	320	312.52	.99999	49
12	349	349	286.48	999	48
13	378	378	264.44	999	47
14	407	407	245.55	999	46
15	.00436	.00436	229.18	.99999	**45**
16	465	465	214.86	999	44
17	495̅	495̅	202.22	999	43
18	524	524	190.98	999	42
19	553	553	180.93	998	41
20	.00582	.00582	171.89	.99998	**40**
21	611	611	163.70	998	39
22	640	640	156.26	998	38
23	669	669	149.47	998	37
24	698	698	143.24	998	36
25	.00727	.00727	137.51	.99997	**35**
26	756	756	132.22	997	34
27	785	785	127.32	997	33
28	814	815̅	122.77	997	32
29	844	844	118.54	996	31
30	.00873	.00873	114.59	.99996	**30**
31	902	902	110.89	996	29
32	931	931	107.43	996	28
33	960	960	104.17	995	27
34	.00989	.00989	101.11	995	26
35	.01018	.01018	98.218	.99995̅	**25**
36	047	047	95.489	995̅	24
37	076	076	92.908	994	23
38	105	105	90.463	994	22
39	134	135̅	88.144	994	21
40	.01164	.01164	85.940	.99993	**20**
41	193	193	83.844	993	19
42	222	222	81.847	993	18
43	251	251	79.943	992	17
44	280	280	78.126	992	16
45	.01309	.01309	76.390	.99991	**15**
46	338	338	74.729	991	14
47	367	367	73.139	991	13
48	396	396	71.615	990	12
49	425	425	70.153	990	11
50	.01454	.01455̅	68.750	.99989	**10**
51	483	484	67.402	989	9
52	513	513	66.105	989	8
53	542	542	64.858	988	7
54	571	571	63.657	988	6
55	.01600	.01600	62.499	.99987	**5**
56	629	629	61.383	987	4
57	658	658	60.306	986	3
58	687	687	59.266	986	2
59	716	716	58.261	985	1
60	.01745	.01746	57.290	.99985̅	**0**
	cos	cot	tan	sin	′

89°

1°

′	sin	tan	cot	cos	
0	.01745	.01746	57.290	.99985̅	**60**
1	774	775̅	56.351	984	59
2	803	804	55.442	984	58
3	832	833	54.561	983	57
4	862	862	53.709	983	56
5	.01891	.01891	52.882	.99982	**55**
6	920	920	52.081	982	54
7	949	949	51.303	981	53
8	.01978	.01978	50.549	980	52
9	.02007	.02007	49.816	980	51
10	.02036	.02036	49.104	.99979	**50**
11	065	066	48.412	979	49
12	094	095̅	47.740	978	48
13	123	124	47.085	977	47
14	152	153	46.449	977	46
15	.02181	.02182	45.829	.99976	**45**
16	211	211	45.226	976	44
17	240	240	44.639	975̅	43
18	269	269	44.066	974	42
19	298	298	43.508	974	41
20	.02327	.02328	42.964	.99973	**40**
21	356	357	42.433	972	39
22	385	386	41.916	972	38
23	414	415̅	41.411	971	37
24	443	444	40.917	970	36
25	.02472	.02473	40.436	.99969	**35**
26	501	502	39.965	969	34
27	530	531	39.506	968	33
28	560	560	39.057	967	32
29	589	589	38.618	966	31
30	.02618	.02619	38.188	.99966	**30**
31	647	648	37.769	965̅	29
32	676	677	37.358	964	28
33	705̅	706	36.956	963	27
34	734	735	36.563	963	26
35	.02763	.02764	36.178	.99962	**25**
36	792	793	35.801	961	24
37	821	822	35.431	960	23
38	850	851	35.070	959	22
39	879	881	34.715	959	21
40	.02908	.02910	34.368	.99958	**20**
41	938	939	34.027	957	19
42	967	968	33.694	956	18
43	.02996	.02997	33.366	955	17
44	.03025̅	.03026	33.045	954	16
45	.03054	.03055	32.730	.99953	**15**
46	083	084	32.421	952	14
47	112	114	32.118	952	13
48	141	143	31.821	951	12
49	170	172	31.528	95̅0	11
50	.03199	.03201	31.242	.99949	**10**
51	228	230	30.960	948	9
52	257	259	30.683	947	8
53	286	288	30.412	946	7
54	316	317	30.145̅	945	6
55	.03345̅	.03346	29.882	.99944	**5**
56	374	376	29.624	943	4
57	403	405̅	29.371	942	3
58	432	434	29.122	941	2
59	461	463	28.877	940	1
60	.03490	.03492	28.636	.99939	**0**
	cos	cot	tan	sin	′

88°

TABLE III

2°

′	sin	tan	cot	cos	
0	.03490	.03492	28.636	.99939	**60**
1	519	521	.399	938	59
2	548	550	28.166	937	58
3	577	579	27.937	936	57
4	606	609	.712	935	56
5	.03635	.03638	27.490	.99934	**55**
6	664	667	.271	933	54
7	693	696	27.057	932	53
8	723	725	26.845	931	52
9	752	754	.637	930	51
10	.03781	.03783	26.432	.99929	**50**
11	810	812	.230	927	49
12	839	842	26.031	926	48
13	868	871	25.835	925	47
14	897	900	.642	924	46
15	.03926	.03929	25.452	.99923	**45**
16	955	958	.264	922	44
17	.03984	.03987	25.080	921	43
18	.04013	.04016	24.898	919	42
19	042	046	.719	918	41
20	.04071	.04075	24.542	.99917	**40**
21	100	104	.368	916	39
22	129	133	.196	915	38
23	159	162	24.026	913	37
24	188	191	23.859	912	36
25	.04217	.04220	23.695	.99911	**35**
26	246	250	.532	910	34
27	275	279	.372	909	33
28	304	308	.214	907	32
29	333	337	23.058	906	31
30	.04362	.04366	22.904	.99905	**30**
31	391	395	.752	904	29
32	420	424	.602	902	28
33	449	454	.454	901	27
34	478	483	.308	900	26
35	.04507	.04512	22.164	.99898	**25**
36	536	541	22.022	897	24
37	565	570	21.881	896	23
38	594	599	.743	894	22
39	623	628	.606	893	21
40	.04653	.04658	21.470	.99892	**20**
41	682	687	.337	890	19
42	711	716	.205	889	18
43	740	745	21.075	888	17
44	769	774	20.946	886	16
45	.04798	.04803	20.819	.99885	**15**
46	827	833	.693	883	14
47	856	862	.569	882	13
48	885	891	.446	881	12
49	914	920	.325	879	11
50	.04943	.04949	20.206	.99878	**10**
51	.04972	.04978	20.087	876	9
52	.05001	.05007	19.970	875	8
53	030	037	.855	873	7
54	059	066	.740	872	6
55	.05088	.05095	19.627	.99870	**5**
56	117	124	.516	869	4
57	146	153	.405	867	3
58	175	182	.296	866	2
59	205	212	.188	864	1
60	.05234	.05241	19.081	.99863	**0**
	cos	cot	tan	sin	′

87°

3°

′	sin	tan	cot	cos	
0	.05234	.05241	19.081	.99863	**60**
1	263	270	18.976	861	59
2	292	299	.871	860	58
3	321	328	.768	858	57
4	350	357	.666	857	56
5	.05379	.05387	18.564	.99855	**55**
6	408	416	.464	854	54
7	437	445	.366	852	53
8	466	474	.268	851	52
9	495	503	.171	849	51
10	.05524	.05533	18.075	.99847	**50**
11	553	562	17.980	846	49
12	582	591	.886	844	48
13	611	620	.793	842	47
14	640	649	.702	841	46
15	.05669	.05678	17.611	.99839	**45**
16	698	708	.521	838	44
17	727	737	.431	836	43
18	756	766	.343	834	42
19	785	795	.256	833	41
20	.05814	.05824	17.169	.99831	**40**
21	844	854	17.084	829	39
22	873	883	16.999	827	38
23	902	912	.915	826	37
24	931	941	.832	824	36
25	.05960	.05970	16.750	.99822	**35**
26	.05989	.05999	.668	821	34
27	.06018	.06029	.587	819	33
28	047	058	.507	817	32
29	076	087	.428	815	31
30	.06105	.06116	16.350	.99813	**30**
31	134	145	.272	812	29
32	163	175	.195	810	28
33	192	204	.119	808	27
34	221	233	16.043	806	26
35	.06250	.06262	15.969	.99804	**25**
36	279	291	.895	803	24
37	308	321	.821	801	23
38	337	350	.748	799	22
39	366	379	.676	797	21
40	.06395	.06408	15.605	.99795	**20**
41	424	438	.534	793	19
42	453	467	.464	792	18
43	482	496	.394	790	17
44	511	525	.325	788	16
45	.06540	.06554	15.257	.99786	**15**
46	569	584	.189	784	14
47	598	613	.122	782	13
48	627	642	15.056	780	12
49	656	671	14.990	778	11
50	.06685	.06700	14.924	.99776	**10**
51	714	730	.860	774	9
52	743	759	.795	772	8
53	773	788	.732	770	7
54	802	817	.669	768	6
55	.06831	.06847	14.606	.99766	**5**
56	860	876	.544	764	4
57	889	905	.482	762	3
58	918	934	.421	760	2
59	947	963	.361	758	1
60	.06976	.06993	14.301	.99756	**0**
	cos	cot	tan	sin	′

86°

′	sin	tan	cot	cos	
0	.06976	.06993	14.301	.99756	**60**
1	.07005̄	.07022	.241	754	59
2	034	051	.182	752	58
3	063	080	.124	750	57
4	092	110	.065	748	56
5	.07121	.07139	14.008	.99746	**55**
6	150	168	13.951	744	54
7	179	197	.894	742	53
8	208	227	.838	740	52
9	237	256	.782	738	51
10	.07266	.07285	13.727	.99736	**50**
11	295̄	314	.672	734	49
12	324	344	.617	731	48
13	353	373	.563	729	47
14	382	402	.510	727	46
15	.07411	.07431	13.457	.99725	**45**
16	440	461	.404	723	44
17	469	490	.352	721	43
18	498	519	.300	719	42
19	527	548	.248	716	41
20	.07556	.07578	13.197	.99714	**40**
21	585̄	607	.146	712	39
22	614	636	.096	710	38
23	643	665	13.046	708	37
24	672	695̄	12.996	705	36
25	.07701	.07724	12.947	.99703	**35**
26	730	753	.898	701	34
27	759	782	.85̄0	699	33
28	788	812	.801	696	32
29	817	841	.754	694	31
30	.07846	.07870	12.706	.99692	**30**
31	875̄	899	.659	689	29
32	904	929	.612	687	28
33	933	958	.566	685̄	27
34	962	.07987	.520	683	26
35	.07991	.08017	12.474	.99680	**25**
36	.08020	046	.429	678	24
37	049	075	.384	676	23
38	078	104	.339	673	22
39	107	134	.295̄	671	21
40	.08136	.08163	12.251	.99668	**20**
41	165̄	192	.207	666	19
42	194	221	.163	664	18
43	223	251	.120	661	17
44	252	280	.077	659	16
45	.08281	.08309	12.035̄	.99657	**15**
46	310	339	11.992	654	14
47	339	368	.950	652	13
48	368	397	.909	649	12
49	397	427	.867	647	11
50	.08426	.08456	11.826	.99644	**10**
51	455̄	485	.785	642	9
52	484	514	.745̄	639	8
53	513	544	.705̄	637	7
54	542	573	.664	635̄	6
55	.08571	.08602	11.625̄	.99632	**5**
56	600	632	.585	630	4
57	629	661	.546	627	3
58	658	690	.507	625̄	2
59	687	720	.468	622	1
60	.08716	.08749	11.430	.99619	**0**
	cos	cot	tan	sin	′

′	sin	tan	cot	cos	
0	.08716	.08749	11.430	.99619	**60**
1	745̄	778	.392	617	59
2	774	807	.354	614	58
3	803	837	.316	612	57
4	831	866	.279	609	56
5	.08860	.08895	11.242	.99607	**55**
6	889	925̄	.205̄	604	54
7	918	954	.168	602	53
8	947	.08983	.132	599	52
9	.08976	.09013	.095	596	51
10	.09005	.09042	11.059	.99594	**50**
11	034	071	11.024	591	49
12	063	101	10.988	588	48
13	092	130	.953	586	47
14	121	159	.918	583	46
15	.09150	.09189	10.883	.99580	**45**
16	179	218	.848	578	44
17	208	247	.814	575	43
18	237	277	.780	572	42
19	266	306	.746	570	41
20	.09295̄	.09335	10.712	.99567	**40**
21	324	365̄	.678	564	39
22	353	394	.645̄	562	38
23	382	423	.612	559	37
24	411	453	.579	556	36
25	.09440	.09482	10.546	.99553	**35**
26	469	511	.514	551	34
27	498	541	.481	548	33
28	527	570	.449	545	32
29	556	600	.417	542	31
30	.09585̄	.09629	10.385	.99540	**30**
31	614	658	.354	537	29
32	642	688	.322	534	28
33	671	717	.291	531	27
34	700	746	.260	528	26
35	.09729	.09776	10.229	.99526	**25**
36	758	805	.199	523	24
37	787	834	.168	520	23
38	816	864	.138	517	22
39	845	893	.108	514	21
40	.09874	.09923	10.078	.99511	**20**
41	903	952	.048	508	19
42	932	.09981	10.019	506	18
43	961	.10011	9.9893	503	17
44	.09990	040	.9601	5̄00	16
45	.10019	.10069	9.9310	.99497	**15**
46	048	099	.9021	494	14
47	077	128	.8734	491	13
48	106	158	.8448	488	12
49	135̄	187	.8164	485	11
50	.10164	.10216	9.7882	.99482	**10**
51	192	246	.7601	479	9
52	221	275	.7322	476	8
53	250	305̄	.7044	473	7
54	279	334	.6768	470	6
55	.10308	.10363	9.6493	.99467	**5**
56	337	393	.6220	464	4
57	366	422	.5949	461	3
58	395̄	452	.5679	458	2
59	424	481	.5411	455	1
60	.10453	.10510	9.5144	.99452	**0**
	cos	cot	tan	sin	′

85° **84°**

6°

′	sin	tan	cot	cos	
0	.10453	.10510	9.5144	.99452	60
1	482	540	.4878	449	59
2	511	569	.4614	446	58
3	540	599	.4352	443	57
4	569	628	.4090	440	56
5	.10597	.10657	9.3831	.99437	55
6	626	687	.3572	434	54
7	655	716	.3315	431	53
8	684	746	.3060	428	52
9	713	775	.2806	424	51
10	.10742	.1080$\bar{5}$	9.2553	.99421	50
11	771	834	.2302	418	49
12	800	863	.2052	415	48
13	829	893	.1803	412	47
14	858	922	.1555	409	46
15	.10887	.10952	9.1309	.99406	45
16	916	.10981	.106$\bar{5}$	402	44
17	94$\bar{5}$	.11011	.0821	399	43
18	.10973	040	.0579	396	42
19	.11002	070	.0338	393	41
20	.11031	.11099	9.0098	.99390	40
21	060	128	8.9860	386	39
22	089	158	.9623	383	38
23	118	187	.9387	380	37
24	147	217	.9152	377	36
25	.11176	.11246	8.8919	.99374	35
26	20$\bar{5}$	276	.8686	370	34
27	234	305	.8455	367	33
28	263	33$\bar{5}$	.8225	364	32
29	291	364	.7996	360	31
30	.11320	.11394	8.7769	.99357	30
31	349	423	.7542	354	29
32	378	452	.7317	351	28
33	407	482	.7093	347	27
34	436	511	.6870	344	26
35	.1146$\bar{5}$	.11541	8.6648	.99341	25
36	494	570	.6427	337	24
37	523	600	.6208	334	23
38	552	629	.5989	331	22
39	580	659	.5772	327	21
40	.11609	.11688	8.5555	.99324	20
41	638	718	.5340	320	19
42	667	747	.5126	317	18
43	696	777	.4913	314	17
44	72$\bar{5}$	806	.4701	310	16
45	.11754	.11836	8.4490	.99307	15
46	783	865	.4280	303	14
47	812	89$\bar{5}$	.4071	300	13
48	840	924	.3863	297	12
49	869	954	.3656	293	11
50	.11898	.11983	8.34$\bar{5}$0	.99290	10
51	927	.12013	.324$\bar{5}$	286	9
52	956	042	.3041	283	8
53	.1198$\bar{5}$	072	.2838	279	7
54	.12014	101	.2636	276	6
55	.12043	.12131	8.2434	.99272	5
56	071	160	.2234	269	4
57	100	190	.2035	265	3
58	129	219	.1837	262	2
59	158	249	.1640	258	1
60	.12187	.12278	8.1443	.9925$\bar{5}$	0
	cos	cot	tan	sin	′

83°

7°

′	sin	tan	cot	cos	
0	.12187	.12278	8.1443	.9925$\bar{5}$	60
1	216	308	.1248	251	59
2	24$\bar{5}$	338	.1054	248	58
3	274	367	.0860	244	57
4	302	397	.0667	240	56
5	.12331	.12426	8.0476	.99237	55
6	360	456	.028$\bar{5}$	233	54
7	389	485	8.009$\bar{5}$	230	53
8	418	51$\bar{5}$	7.9906	226	52
9	447	544	.9718	222	51
10	.12476	.12574	7.9530	.99219	50
11	504	603	.9344	215	49
12	533	633	.9158	211	48
13	562	662	.8973	208	47
14	591	692	.8789	204	46
15	.12620	.12722	7.8606	.99200	45
16	649	751	8424	197	44
17	678	781	.8243	193	43
18	706	810	.8062	189	42
19	735	840	.7882	186	41
20	.12764	.12869	7.7704	.99182	40
21	793	899	.7525	178	39
22	822	929	.7348	17$\bar{5}$	38
23	851	958	.7171	171	37
24	880	.12988	.6996	167	36
25	.12908	.13017	7.6821	.99163	35
26	937	047	.6647	160	34
27	966	076	.6473	156	33
28	.1299$\bar{5}$	106	.6301	152	32
29	.13024	136	.6129	148	31
30	.13053	.13165	7.5958	.99144	30
31	081	19$\bar{5}$	.5787	141	29
32	110	224	.5618	137	28
33	139	254	.5449	133	27
34	168	284	.5281	129	26
35	.13197	.13313	7.5113	.99125	25
36	226	343	.4947	122	24
37	254	372	.4781	118	23
38	283	402	.4615	114	22
39	312	432	.4451	110	21
40	.13341	.13461	7.4287	.99106	20
41	370	491	.4124	102	19
42	399	521	.3962	098	18
43	427	550	.3800	094	17
44	456	580	.3639	091	16
45	.13485	.13609	7.3479	.99087	15
46	514	639	.3319	083	14
47	543	669	.3160	079	13
48	572	698	.3002	07$\bar{5}$	12
49	600	728	.2844	071	11
50	.13629	.13758	7.2687	.99067	10
51	658	787	.2531	063	9
52	687	817	.2375	059	8
53	716	846	.2220	05$\bar{5}$	7
54	744	876	.2066	051	6
55	.13773	.13906	7.1912	.99047	5
56	802	935	.1759	043	4
57	831	965	.1607	039	3
58	860	.1399$\bar{5}$	.1455	03$\bar{5}$	2
59	889	.14024	.1304	031	1
60	.13917	.14054	7.1154	.99027	0
	cos	cot	tan	sin	′

82°

8°

′	sin	tan	cot	cos	
0	.13917	.14054	7.1154	.99027	**60**
1	946	084	.1004	023	59
2	.13975̄	113	.0855̄	019	58
3	.14004	143	.0706	015̄	57
4	033	173	.0558	011	56
5	.14061	.14202	7.0410	.99006	**55**
6	090	232	.0264	.99002	54
7	119	262	7.0117	.98998	53
8	148	291	6.9972	994	52
9	177	321	.9827	990	51
10	.14205	.14351	6.9682	.98986	**50**
11	234	381	.9538	982	49
12	263	410	.9395	978	48
13	292	440	.9252	973	47
14	320	470	.9110	969	46
15	.14349	.14499	6.8969	.98965	**45**
16	378	529	.8828	961	44
17	407	559	.8687	957	43
18	436	588	.8548	953	42
19	464	618	.8408	948	41
20	.14493	.14648	6.8269	.98944	**40**
21	522	678	.8131	940	39
22	551	707	.7994	936	38
23	580	737	.7856	931	37
24	608	767	.7720	927	36
25	.14637	.14796	6.7584	.98923	**35**
26	666	826	.7448	919	34
27	695̄	856	.7313	914	33
28	723	886	.7179	910	32
29	752	915	.7045̄	906	31
30	.14781	.14945	6.6912	.98902	**30**
31	810	.14975̄	.6779	897	29
32	838	.15005̄	.6646	893	28
33	867	034	.6514	889	27
34	896	064	.6383	884	26
35	.14925̄	.15094	6.6252	.98880	**25**
36	954	124	.6122	876	24
37	.14982	153	.5992	871	23
38	.15011	183	.5863	867	22
39	040	213	.5734	863	21
40	.15069	.15243	6.5606	.98858	**20**
41	097	272	.5478	854	19
42	126	302	.5350	849	18
43	155̄	332	.5223	845̄	17
44	184	362	.5097	841	16
45	.15212	.15391	6.4971	.98836	**15**
46	241	421	.4846	832	14
47	270	451	.4721	827	13
48	299	481	.4596	823	12
49	327	511	.4472	818	11
50	.15356	.15540	6.4348	.98814	**10**
51	385̄	570	.4225	809	9
52	414	600	.4103	805̄	8
53	442	630	.3980	800	7
54	471	660	.3859	796	6
55	.15̄500	.15689	6.3737	.98791	**5**
56	529	719	.3617	787	4
57	557	749	.3496	782	3
58	586	779	.3376	778	2
59	615̄	809	.3257	773	1
60	.15643	.15838	6.3138	.98769	**0**
	cos	cot	tan	sin	′

81°

9°

′	sin	tan	cot	cos	
0	.15643	.15838	6.3138	.98769	**60**
1	672	868	.3019	764	59
2	701	898	.2901	760	58
3	730	928	.2783	755	57
4	758	958	.2666	751	56
5	.15787	.15988	6.2549	.98746	**55**
6	816	.16017	.2432	741	54
7	845̄	047	.2316	737	53
8	873	077	.2200	732	52
9	902	107	.2085	728	51
10	.15931	.16137	6.1970	.98723	**50**
11	959	167	.1856	718	49
12	.15988	196	.1742	714	48
13	.16017	226	.1628	709	47
14	046	256	.1515	704	46
15	.16074	.16286	6.1402	.98700	**45**
16	103	316	.1290	695̄	44
17	132	346	.1178	690	43
18	160	376	.1066	686	42
19	189	405	.0955	681	41
20	.16218	.16435	6.0844	.98676	**40**
21	246	465	.0734	671	39
22	275	495	.0624	667	38
23	304	525	.0514	662	37
24	333	555̄	.0405	657	36
25	.16361	.16585̄	6.0296	.98652	**35**
26	390	615̄	.0188	648	34
27	419	645̄	6.0080	643	33
28	447	674	5.9972	638	32
29	476	704	.9865̄	633	31
30	.16505̄	.16734	5.9758	.98629	**30**
31	533	764	.9651	624	29
32	562	794	.9545̄	619	28
33	591	824	.9439	614	27
34	620	854	.9333	609	26
35	.16648	.16884	5.9228	.98604	**25**
36	677	914	.9124	600	24
37	706	944	.9019	595̄	23
38	734	.16974	8915	590	22
39	763	.17004	.8811	585	21
40	.16792	.17033	5.8708	.98580	**20**
41	820	063	.8605	575	19
42	849	093	.8502	570	18
43	878	123	.8400	565	17
44	906	153	.8298	561	16
45	.16935̄	.17183	5.8197	.98556	**15**
46	964	213	.8095	551	14
47	16992	243	.7994	546	13
48	.17021	273	.7894	541	12
49	0̄50	303	.7794	536	11
50	.17078	.17333	5.7694	.98531	**10**
51	107	363	.7594	526	9
52	136	393	.7495̄	521	8
53	164	423	.7396	516	7
54	193	453	.7297	511	6
55	.17222	.17483	5.7199	.98506	**5**
56	250	513	.7101	501	4
57	279	543	.7004	496	3
58	308	573	.6906	491	2
59	336	603	.6809	486	1
60	.17365̄	.17633	5.6713	.98481	**0**
	cos	cot	tan	sin	′

80°

10°

′	sin	tan	cot	cos	
0	.17365	.17633	5.6713	.98481	60
1	393	663	.6617	476	59
2	422	693	.6521	471	58
3	451	723	.6425̄	466	57
4	479	753	.6329	461	56
5	.17508	.17783	5.6234	.98455	55
6	537	813	.6140	450	54
7	565	843	.6045	445	53
8	594	873	.5951	440	52
9	623	903	.5857	435̄	51
10	.17651	.17933	5.5764	.98430	50
11	680	963	.5671	425̄	49
12	708	.17993	.5578	420	48
13	737	.18023	.5485	414	47
14	766	053	.5393	409	46
15	.17794	.18083	5.5301	.98404	45
16	823	113	.5209	399	44
17	852	143	.5118	394	43
18	880	173	.5026	389	42
19	909	203	.4936	383	41
20	.17937	.18233	5.4845	.98378	40
21	966	263	.4755̄	373	39
22	.17995̄	293	.4665̄	368	38
23	.18023	323	.4575	362	37
24	052	353	.4486	357	36
25	.18081	.18384	5.4397	.98352	35
26	109	414	.4308	347	34
27	138	444	.4219	341	33
28	166	474	.4131	336	32
29	195̄	504	.4043	331	31
30	.18224	.18534	5.3955	.98325	30
31	252	564	.3868	320	29
32	281	594	.3781	315̄	28
33	309	624	.3694	310	27
34	338	654	.3607	304	26
35	.18367	.18684	5.3521	.98299	25
36	395	714	.3435̄	294	24
37	424	745̄	.3349	288	23
38	452	775̄	.3263	283	22
39	481	805̄	.3178	277	21
40	.18509	.18835̄	5.3093	.98272	20
41	538	865	.3008	267	19
42	567	895	.2924	261	18
43	595	925	.2839	256	17
44	624	955	.2755	250	16
45	.18652	.18986	5.2672	.98245	15
46	681	.19016	.2588	240	14
47	710	046	.2505̄	234	13
48	738	076	.2422	229	12
49	767	106	.2339	223	11
50	.18795	.19136	5.2257	.98218	10
51	824	166	.2174	212	9
52	852	197	.2092	207	8
53	881	227	.2011	201	7
54	910	257	.1929	196	6
55	.18938	.19287	5.1848	.98190	5
56	967	317	.1767	185̄	4
57	.18995	347	.1686	179	3
58	.19024	378	.1606	174	2
59	052	408	.1526	168	1
60	.19081	.19438	5.1446	.98163	0
	cos	cot	tan	sin	′

79°

11°

′	sin	tan	cot	cos	
0	.19081	.19438	5.1446	.98163	60
1	109	468	.1366	157	59
2	138	498	.1286	152	58
3	167	529	.1207	146	57
4	195	559	.1128	140	56
5	.19224	.19589	5.1049	.98135̄	55
6	252	619	.0970	129	54
7	281	649	.0892	124	53
8	309	680	.0814	118	52
9	338	710	.0736	112	51
10	.19366	.19740	5.0658	.98107	50
11	395̄	770	.0581	101	49
12	423	801	.0504	096	48
13	452	831	.0427	090	47
14	481	861	.035̄0	084	46
15	.19509	.19891	5.0273	.98079	45
16	538	921	.0197	073	44
17	566	952	.0121	067	43
18	595̄	.19982	5.0045	061	42
19	623	.20012	4.9969	056	41
20	.19652	.20042	4.9894	.98050	40
21	680	073	.9819	044	39
22	709	103	.9744	039	38
23	737	133	.9669	033	37
24	766	164	.9594	027	36
25	.19794	.20194	4.9520	.98021	35
26	823	224	.9446	016	34
27	851	254	.9372	010	33
28	880	285̄	.9298	.98004	32
29	908	315̄	.9225̄	.97998	31
30	.19937	.20345	4.9152	.97992	30
31	965	376	.9078	987	29
32	.19994	406	.9006	981	28
33	.20022	436	.8933	975	27
34	051	466	.8860	969	26
35	.20079	.20497	4.8788	.97963	25
36	108	527	.8716	958	24
37	136	557	.8644	952	23
38	165̄	588	.8573	946	22
39	193	618	.8501	940	21
40	.20222	.20648	4.8430	.97934	20
41	250	679	.8359	928	19
42	279	709	.8288	922	18
43	307	739	.8218	916	17
44	336	770	.8147	910	16
45	.20364	.20800	4.8077	.97905̄	15
46	393	830	.8007	899	14
47	421	861	.7937	893	13
48	45̄0	891	.7867	887	12
49	478	921	.7798	881	11
50	.20507	.20952	4.7729	.97875̄	10
51	535	.20982	.7659	869	9
52	563	.21013	.7591	863	8
53	592	043	.7522	857	7
54	620	073	.7453	851	6
55	.20649	.21104	4.7385	.97845̄	5
56	677	134	.7317	839	4
57	706	164	.7249	833	3
58	734	195̄	.7181	827	2
59	763	225	.7114	821	1
60	.20791	.21256	4.7046	.97815̄	0
	cos	cot	tan	sin	′

78°

12°

′	sin	tan	cot	cos	
0	.20791	.21256	4.7046	.97815̄	**60**
1	820	286	.6979	809	59
2	848	316	.6912	803	58
3	877	347	.6845	797	57
4	905̄	377	.6779	791	56
5	.20933	.21408	4.6712	.97784	**55**
6	962	438	.6646	778	54
7	.20990	469	.6580	772	53
8	.21019	499	.6514	766	52
9	047	529	.6448	760	51
10	.21076	.21560	4.6382	.97754	**50**
11	104	590	.6317	748	49
12	132	621	.6252	742	48
13	161	651	.6187	735	47
14	189	682	.6122	729	46
15	.21218	.21712	4.6057	.97723	**45**
16	246	743	.5993	717	44
17	275̄	773	.5928	711	43
18	303	804	.5864	705̄	42
19	331	834	.5800	698	41
20	.21360	.21864	4.5736	.97692	**40**
21	388	895̄	.5673	686	39
22	417	925	.5609	680	38
23	445	956	.5546	673	37
24	474	.21986	.5483	667	36
25	.21502	.22017	4.5420	.97661	**35**
26	530	047	.5357	655̄	34
27	559	078	.5294	648	33
28	587	108	.5232	642	32
29	616	139	.5169	636	31
30	.21644	.22169	4.5107	.97630	**30**
31	672	200	.5045	623	29
32	701	231	.4983	617	28
33	729	261	.4922	611	27
34	758	292	.4860	604	26
35	.21786	.22322	4.4799	.97598	**25**
36	814	353	.4737	592	24
37	843	383	.4676	585	23
38	871	414	.4615	579	22
39	899	444	.4555̄	573	21
40	.21928	.22475̄	4.4494	.97566	**20**
41	956	505	.4434	560	19
42	.21985̄	536	.4373	553	18
43	.22013	567	.4313	547	17
44	041	597	.4253	541	16
45	.22070	.22628	4.4194	.97534	**15**
46	098	658	.4134	528	14
47	126	689	.4075̄	521	13
48	155̄	719	.4015	515̄	12
49	183	750	.3956	508	11
50	.22212	.22781	4.3897	.97502	**10**
51	240	811	.3838	496	9
52	268	842	.3779	489	8
53	297	872	.3721	483	7
54	325	903	.3662	476	6
55	.22353	.22934	4.3604	.97470	**5**
56	382	964	.3546	463	4
57	410	.22995̄	.3488	457	3
58	438	.23026	.3430	450	2
59	467	056	.3372	444	1
60	.22495	.23087	4.3315̄	.97437	**0**
	cos	cot	tan	sin	′

77°

13°

′	sin	tan	cot	cos	
0	.22495	.23087	4.3315̄	.97437	**60**
1	523	117	.3257	430	59
2	552	148	.3200	424	58
3	580	179	.3143	417	57
4	608	209	.3086	411	56
5	.22637	.23240	4.3029	.97404	**55**
6	665	271	.2972	398	54
7	693	301	.2916	391	53
8	722	332	.2859	384	52
9	750	363	.2803	378	51
10	.22778	.23393	4.2747	.97371	**50**
11	807	424	.2691	365̄	49
12	835	455̄	.2635	358	48
13	863	485	.2580	351	47
14	892	516	.2524	345̄	46
15	.22920	.23547	4.2468	.97338	**45**
16	948	578	.2413	331	44
17	.22977	608	.2358	325̄	43
18	.23005̄	639	.2303	318	42
19	033	670	.2248	311	41
20	.23062	.23700	4.2193	.97304	**40**
21	090	731	.2139	298	39
22	118	762	.2084	291	38
23	146	793	.2030	284	37
24	175̄	823	.1976	278	36
25	.23203	.23854	4.1922	.97271	**35**
26	231	885̄	.1868	264	34
27	260	916	.1814	257	33
28	288	946	.1760	251	32
29	316	.23977	.1706	244	31
30	.23345̄	.24008	4.1653	.97237	**30**
31	373	039	.1600	230	29
32	401	069	.1547	223	28
33	429	100	.1493	217	27
34	458	131	.1441	210	26
35	.23486	.24162	4.1388	.97203	**25**
36	514	193	.1335	196	24
37	542	223	.1282	189	23
38	571	254	.1230	182	22
39	599	285̄	.1178	176	21
40	.23627	.24316	4.1126	.97169	**20**
41	656	347	.1074	162	19
42	684	377	.1022	155̄	18
43	712	408	.0970	148	17
44	740	439	.0918	141	16
45	.23769	.24470	4.0867	.97134	**15**
46	797	501	.0815	127	14
47	825	532	.0764	120	13
48	853	562	.0713	113	12
49	882	593	.0662	106	11
50	.23910	.24624	4.0611	.97100	**10**
51	938	655̄	.0560	093	9
52	966	686	.0509	086	8
53	.23995̄	717	.0459	079	7
54	.24023	747	.0408	072	6
55	.24051	.24778	4.0358	.97065̄	**5**
56	079	809	.0308	058	4
57	108	840	.0257	051	3
58	136	871	.0207	044	2
59	164	902	.0158	037	1
60	.24192	.24933	4.0108	.97030	**0**
	cos	cot	tan	sin	′

76°

14°

′	sin	tan	cot	cos	
0	.24192	.24933	4.0108	.97030	60
1	220	964	.0058	023	59
2	249	.24995	4.0009	015	58
3	277	.25026	3.9959	008	57
4	305	056	.9910	.97001	56
5	.24333	.25087	3.9861	.96994	55
6	362	118	.9812	987	54
7	390	149	.9763	980	53
8	418	180	.9714	973	52
9	446	211	.9665	966	51
10	.24474	.25242	3.9617	.96959	50
11	503	273	.9568	952	49
12	531	304	.9520	945	48
13	559	335	.9471	937	47
14	587	366	.9423	930	46
15	.24615	.25397	3.9375	.96923	45
16	644	428	.9327	916	44
17	672	459	.9279	909	43
18	700	490	.9232	902	42
19	728	521	.9184	894	41
20	.24756	.25552	3.9136	.96887	40
21	784	583	.9089	880	39
22	813	614	.9042	873	38
23	841	645	.8995	866	37
24	869	676	.8947	858	36
25	.24897	.25707	3.8900	.96851	35
26	925	738	.8854	844	34
27	954	769	.8807	837	33
28	.24982	800	.8760	829	32
29	.25010	831	.8714	822	31
30	.25038	.25862	3.8667	.96815	30
31	066	893	.8621	807	29
32	094	924	.8575	800	28
33	122	955	.8528	793	27
34	151	.25986	.8482	786	26
35	.25179	.26017	3.8436	.96778	25
36	207	048	.8391	771	24
37	235	079	.8345	764	23
38	263	110	.8299	756	22
39	291	141	.8254	749	21
40	.25320	.26172	3.8208	.96742	20
41	348	203	.8163	734	19
42	376	235	.8118	727	18
43	404	266	.8073	719	17
44	432	297	.8028	712	16
45	.25460	.26328	3.7983	.96705	15
46	488	359	.7938	697	14
47	516	390	.7893	690	13
48	545	421	.7848	682	12
49	573	452	.7804	675	11
50	.25601	.26483	3.7760	.96667	10
51	629	515	.7715	660	9
52	657	546	.7671	653	8
53	685	577	.7627	645	7
54	713	608	.7583	638	6
55	.25741	.26639	3.7539	.96630	5
56	769	670	.7495	623	4
57	798	701	.7451	615	3
58	826	733	.7408	608	2
59	854	764	.7364	600	1
60	.25882	.26795	3.7321	.96593	0
	cos	cot	tan	sin	′

75°

15°

′	sin	tan	cot	cos	
0	.25882	.26795	3.7321	.96593	60
1	910	826	.7277	585	59
2	938	857	.7234	578	58
3	966	888	.7191	570	57
4	.25994	920	.7148	562	56
5	.26022	.26951	3.7105	.96555	55
6	050	.26982	.7062	547	54
7	079	.27013	.7019	540	53
8	107	044	.6976	532	52
9	135	076	.6933	524	51
10	.26163	.27107	3.6891	.96517	50
11	191	138	.6848	509	49
12	219	169	.6806	502	48
13	247	201	.6764	494	47
14	275	232	.6722	486	46
15	.26303	.27263	3.6680	.96479	45
16	331	294	.6638	471	44
17	359	326	.6596	463	43
18	387	357	.6554	456	42
19	415	388	.6512	448	41
20	.26443	.27419	3.6470	.96440	40
21	471	451	.6429	433	39
22	500	482	.6387	425	38
23	528	513	.6346	417	37
24	556	545	.6305	410	36
25	.26584	.27576	3.6264	.96402	35
26	612	607	.6222	394	34
27	640	638	.6181	386	33
28	668	670	.6140	379	32
29	696	701	.6100	371	31
30	.26724	.27732	3.6059	.96363	30
31	752	764	.6018	355	29
32	780	795	.5978	347	28
33	808	826	.5937	340	27
34	836	858	.5897	332	26
35	.26864	.27889	3.5856	.96324	25
36	892	921	.5816	316	24
37	920	952	.5776	308	23
38	948	.27983	.5736	301	22
39	.26976	.28015	.5696	293	21
40	.27004	.28046	3.5656	.96285	20
41	032	077	.5616	277	19
42	060	109	.5576	269	18
43	088	140	.5536	261	17
44	116	172	.5497	253	16
45	.27144	.28203	3.5457	.96246	15
46	172	234	.5418	238	14
47	200	266	.5379	230	13
48	228	297	.5339	222	12
49	256	329	.5300	214	11
50	.27284	.28360	3.5261	.96206	10
51	312	391	.5222	198	9
52	340	423	.5183	190	8
53	368	454	.5144	182	7
54	396	486	.5105	174	6
55	.27424	.28517	3.5067	.96166	5
56	452	549	.5028	158	4
57	480	580	.4989	150	3
58	508	612	.4951	142	2
59	536	643	.4912	134	1
60	.27564	.28675	3.4874	.96126	0
	cos	cot	tan	sin	′

74°

16°

′	sin	tan	cot	cos	
0	.27564	.28675	3.4874	.96126	60
1	592	706	.4836	118	59
2	620	738	.4798	110	58
3	648	769	.4760	102	57
4	676	801	.4722	094	56
5	.27704	.28832	3.4684	.96086	**55**
6	731	864	.4646	078	54
7	759	895	.4608	070	53
8	787	927	.4570	062	52
9	815	958	.4533	054	51
10	.27843	.28990	3.4495	.96046	**50**
11	871	.29021	.4458	037	49
12	899	053	.4420	029	48
13	927	084	.4383	021	47
14	955	116	.4346	013	46
15	.27983	.29147	3.4308	.96005	**45**
16	.28011	179	.4271	.95997	44
17	039	210	.4234	989	43
18	067	242	.4197	981	42
19	095	274	.4160	972	41
20	.28123	.29305	3.4124	.95964	**40**
21	150	337	.4087	956	39
22	178	368	.4050	948	38
23	206	400	.4014	940	37
24	234	432	.3977	931	36
25	.28262	.29463	3.3941	.95923	**35**
26	290	495	.3904	915	34
27	318	526	.3868	907	33
28	346	558	.3832	898	32
29	374	590	.3796	890	31
30	.28402	.29621	3.3759	.95882	**30**
31	429	653	.3723	874	29
32	457	685	.3687	865	28
33	485	716	.3652	857	27
34	513	748	.3616	849	26
35	.28541	.29780	3.3580	.95841	**25**
36	569	811	.3544	832	24
37	597	843	.3509	824	23
38	625	875	.3473	816	22
39	652	906	.3438	807	21
40	.28680	.29938	3.3402	.95799	**20**
41	708	.29970	.3367	791	19
42	736	.30001	.3332	782	18
43	764	033	.3297	774	17
44	792	065	.3261	766	16
45	.28820	.30097	3.3226	.95757	**15**
46	847	128	.3191	749	14
47	875	160	.3156	740	13
48	903	192	.3122	732	12
49	931	224	.3087	724	11
50	.28959	.30255	3.3052	.95715	**10**
51	.28987	287	.3017	707	9
52	.29015	319	.2983	698	8
53	042	351	.2948	690	7
54	070	382	.2914	681	6
55	.29098	.30414	3.2879	.95673	**5**
56	126	446	.2845	664	4
57	154	478	.2811	656	3
58	182	509	.2777	647	2
59	209	541	.2743	639	1
60	.29237	.30573	3.2709	.95630	**0**
	cos	cot	tan	sin	′

73°

17°

′	sin	tan	cot	cos	
0	.29237	.30573	3.2709	.95630	60
1	265	605	.2675	622	59
2	293	637	.2641	613	58
3	321	669	.2607	605	57
4	348	700	.2573	596	56
5	.29376	.30732	3.2539	.95588	**55**
6	404	764	.2506	579	54
7	432	796	.2472	571	53
8	460	828	.2438	562	52
9	487	860	.2405	554	51
10	.29515	.30891	3.2371	.95545	**50**
11	543	923	.2338	536	49
12	571	955	.2305	528	48
13	599	.30987	.2272	519	47
14	626	.31019	.2238	511	46
15	.29654	.31051	3.2205	.95502	**45**
16	682	083	.2172	493	44
17	710	115	.2139	485	43
18	737	147	.2106	476	42
19	765	178	.2073	467	41
20	.29793	.31210	3.2041	.95459	**40**
21	821	242	.2008	450	39
22	849	274	.1975	441	38
23	876	306	.1943	433	37
24	904	338	.1910	424	36
25	.29932	.31370	3.1878	.95415	**35**
26	960	402	.1845	407	34
27	.29987	434	.1813	398	33
28	.30015	466	.1780	389	32
29	043	498	.1748	380	31
30	.30071	.31530	3.1716	.95372	**30**
31	098	562	.1684	363	29
32	126	594	.1652	354	28
33	154	626	.1620	345	27
34	182	658	.1588	337	26
35	.30209	.31690	3.1556	.95328	**25**
36	237	722	.1524	319	24
37	265	754	.1492	310	23
38	292	786	.1460	301	22
39	320	818	.1429	293	21
40	.30348	.31850	3.1397	.95284	**20**
41	376	882	.1366	275	19
42	403	914	.1334	266	18
43	431	946	.1303	257	17
44	459	.31978	.1271	248	16
45	.30486	.32010	3.1240	.95240	**15**
46	514	042	.1209	231	14
47	542	074	.1178	222	13
48	570	106	.1146	213	12
49	597	139	.1115	204	11
50	.30625	.32171	3.1084	.95195	**10**
51	653	203	.1053	186	9
52	680	235	.1022	177	8
53	708	267	.0991	168	7
54	736	299	.0961	159	6
55	.30763	.32331	3.0930	.95150	**5**
56	791	363	.0899	142	4
57	819	396	.0868	133	3
58	846	428	.0838	124	2
59	874	460	.0807	115	1
60	.30902	.32492	3.0777	.95106	**0**
	cos	cot	tan	sin	′

72°

18°

′	sin	tan	cot	cos	
0	.30902	.32492	3.0777	.95106	60
1	929	524	.0746	097	59
2	957	556	.0716	088	58
3	.30985	588	.0686	079	57
4	.31012	621	.0655	070	56
5	.31040	.32653	3.0625	.95061	55
6	068	685	.0595	052	54
7	095	717	.0565	043	53
8	123	749	.0535	033	52
9	151	782	.0505	024	51
10	.31178	.32814	3.0475	.95015	50
11	206	846	.0445	.95006	49
12	233	878	.0415	.94997	48
13	261	911	.0385	988	47
14	289	943	.0356	979	46
15	.31316	.32975	3.0326	.94970	45
16	344	.33007	.0296	961	44
17	372	040	.0267	952	43
18	399	072	.0237	943	42
19	427	104	.0208	933	41
20	.31454	.33136	3.0178	.94924	40
21	482	169	.0149	915	39
22	510	201	.0120	906	38
23	537	233	.0090	897	37
24	565	266	.0061	888	36
25	.31593	.33298	3.0032	.94878	35
26	620	330	3.0003	869	34
27	648	363	2.9974	860	33
28	675	395	.9945	851	32
29	703	427	.9916	842	31
30	.31730	.33460	2.9887	.94832	30
31	758	492	.9858	823	29
32	786	524	.9829	814	28
33	813	557	.9800	805	27
34	841	589	.9772	795	26
35	.31868	.33621	2.9743	.94786	25
36	896	654	.9714	777	24
37	923	686	.9686	768	23
38	951	718	.9657	758	22
39	.31979	751	.9629	749	21
40	.32006	33783	2.9600	.94740	20
41	034	816	.9572	730	19
42	061	848	.9544	721	18
43	089	881	.9515	712	17
44	116	913	.9487	702	16
45	.32144	.33945	2.9459	.94693	15
46	171	.33978	.9431	684	14
47	199	.34010	.9403	674	13
48	227	043	.9375	665	12
49	254	075	.9347	656	11
50	.32282	.34108	2.9319	.94646	10
51	309	140	.9291	637	9
52	337	173	.9263	627	8
53	364	205	.9235	618	7
54	392	238	.9208	609	6
55	.32419	.34270	2.9180	.94599	5
56	447	303	.9152	590	4
57	474	335	.9125	580	3
58	502	368	.9097	571	2
59	529	400	.9070	561	1
60	.32557	.34433	2.9042	.94552	0
	cos	cot	tan	sin	′

71°

19°

′	sin	tan	cot	cos	
0	.32557	.34433	2.9042	.94552	60
1	584	465	.9015	542	59
2	612	498	.8987	533	58
3	639	530	.8960	523	57
4	667	563	.8933	514	56
5	.32694	.34596	2.8905	.94504	55
6	722	628	.8878	495	54
7	749	661	.8851	485	53
8	777	693	.8824	476	52
9	804	726	.8797	466	51
10	.32832	.34758	2.8770	.94457	50
11	859	791	.8743	447	49
12	887	824	.8716	438	48
13	914	856	.8689	428	47
14	942	889	.8662	418	46
15	.32969	.34922	2.8636	.94409	45
16	.32997	954	.8609	399	44
17	.33024	.34987	.8582	390	43
18	051	.35020	.8556	380	42
19	079	052	.8529	370	41
20	.33106	.35085	2.8502	.94361	40
21	134	118	.8476	351	39
22	161	150	.8449	342	38
23	189	183	.8423	332	37
24	216	216	.8397	322	36
25	.33244	.35248	2.8370	.94313	35
26	271	281	.8344	303	34
27	298	314	.8318	293	33
28	326	346	.8291	284	32
29	353	379	.8265	274	31
30	.33381	.35412	2.8239	.94264	30
31	408	445	.8213	254	29
32	436	477	.8187	245	28
33	463	510	.8161	235	27
34	490	543	.8135	225	26
35	.33518	.35576	2.8109	.94215	25
36	545	608	.8083	206	24
37	573	641	.8057	196	23
38	600	674	.8032	186	22
39	627	707	.8006	176	21
40	.33655	.35740	2.7980	.94167	20
41	682	772	.7955	157	19
42	710	805	.7929	147	18
43	737	838	.7903	137	17
44	764	871	.7878	127	16
45	.33792	.35904	2.7852	.94118	15
46	819	937	.7827	108	14
47	846	.35969	.7801	098	13
48	874	.36002	.7776	088	12
49	901	035	.7751	078	11
50	.33929	.36068	2.7725	.94068	10
51	956	101	.7700	058	9
52	.33983	134	.7675	049	8
53	.34011	167	.7650	039	7
54	038	199	.7625	029	6
55	.34065	.36232	2.7600	.94019	5
56	093	265	.7575	.94009	4
57	120	298	.7550	.93999	3
58	147	331	.7525	989	2
59	175	364	.7500	979	1
60	.34202	.36397	2.7475	.93969	0
	cos	cot	tan	sin	′

70°

20°

′	sin	tan	cot	cos	
0	.34202	.36397	2.747$\bar{5}$	.93969	60
1	229	430	.74$\bar{5}$0	959	59
2	257	463	.7425	949	58
3	284	496	.7400	939	57
4	311	529	.7376	929	56
5	.34339	.36562	2.7351	.93919	55
6	366	59$\bar{5}$	.7326	909	54
7	393	628	.7302	899	53
8	421	661	.7277	889	52
9	448	694	.7253	879	51
10	.34475	.36727	2.7228	.93869	50
11	503	760	.7204	859	49
12	530	793	.7179	849	48
13	557	826	.715$\bar{5}$	839	47
14	584	859	.7130	829	46
15	.34612	.36892	2.7106	.93819	45
16	639	92$\bar{5}$	.7082	809	44
17	666	958	.7058	799	43
18	694	.36991	.7034	789	42
19	721	.37024	.7009	779	41
20	.34748	.37057	2.6985	.93769	40
21	775	090	.6961	759	39
22	803	123	.6937	748	38
23	830	157	.6913	738	37
24	857	190	.6889	728	36
25	.34884	.37223	2.6865	.93718	35
26	912	256	.6841	708	34
27	939	289	.6818	698	33
28	966	322	.6794	688	32
29	.34993	355	.6770	677	31
30	.35021	.37388	2.6746	.93667	30
31	048	422	.6723	657	29
32	075	45$\bar{5}$	.6699	647	28
33	102	488	.6675	637	27
34	130	521	.6652	626	26
35	.35157	.37554	2.6628	.93616	25
36	184	588	.660$\bar{5}$	606	24
37	211	621	.6581	596	23
38	239	654	.6558	585	22
39	266	687	.6534	575	21
40	.35293	.37720	2.6511	.9356$\bar{5}$	20
41	320	754	.6488	55$\bar{5}$	19
42	347	787	.6464	544	18
43	37$\bar{5}$	820	.6441	534	17
44	402	853	.6418	524	16
45	.35429	.37887	2.639$\bar{5}$	.93514	15
46	456	920	.6371	503	14
47	484	953	.6348	493	13
48	511	.37986	.6325	483	12
49	538	.38020	.6302	472	11
50	.35565	.38053	2.6279	.93462	10
51	592	086	.6256	452	9
52	619	120	.6233	441	8
53	647	153	.6210	431	7
54	674	186	.6187	420	6
55	.35701	.38220	2.616$\bar{5}$	.93410	5
56	728	253	.6142	400	4
57	755	286	.6119	389	3
58	782	320	.6096	379	2
59	810	353	.6074	368	1
60	.35837	.38386	2.6051	.93358	0
	cos	cot	tan	sin	′

69°

21°

′	sin	tan	cot	cos	
0	.35837	.38386	2.6051	.93358	60
1	864	420	.6028	348	59
2	891	453	.6006	337	58
3	918	487	.5983	327	57
4	945	520	.5961	316	56
5	.35973	.38553	2.5938	.93306	55
6	.36000	587	.5916	295	54
7	027	620	.5893	28$\bar{5}$	53
8	054	654	.5871	274	52
9	081	687	.5848	264	51
10	.36108	.38721	2.5826	.93253	50
11	135	754	.5804	243	49
12	162	787	.5782	232	48
13	190	821	.5759	222	47
14	217	854	.5737	211	46
15	.36244	.38888	2.571$\bar{5}$	.93201	45
16	271	921	.5693	190	44
17	298	95$\bar{5}$	.5671	180	43
18	325	.38988	.5649	169	42
19	352	.39022	.5627	159	41
20	.36379	.39055	2.560$\bar{5}$	.93148	40
21	406	089	.5583	137	39
22	434	122	.5561	127	38
23	461	156	.5539	116	37
24	488	190	.5517	106	36
25	.3651$\bar{5}$	.39223	2.5495	.9309$\bar{5}$	35
26	542	257	.5473	084	34
27	569	290	.5452	074	33
28	596	324	.5430	063	32
29	623	357	.5408	052	31
30	.36650	.39391	2.5386	.93042	30
31	677	42$\bar{5}$	.536$\bar{5}$	031	29
32	704	458	.5343	020	28
33	731	492	.5322	.93010	27
34	758	526	.5300	.92999	26
35	.36785	.39559	2.5279	.92988	25
36	812	593	.5257	978	24
37	839	626	.5236	967	23
38	867	660	.5214	956	22
39	894	694	.5193	945	21
40	.36921	.39727	2.5172	.9293$\bar{5}$	20
41	948	761	.5150	924	19
42	.3697$\bar{5}$	79$\bar{5}$	.5129	913	18
43	.37002	829	.5108	902	17
44	029	862	.5086	892	16
45	.37056	.39896	2.5065	.92881	15
46	083	930	.5044	870	14
47	110	963	.5023	859	13
48	137	.39997	.5002	849	12
49	164	.40031	.4981	838	11
50	.37191	.4006$\bar{5}$	2.4960	.92827	10
51	218	098	.4939	816	9
52	24$\bar{5}$	132	.4918	805	8
53	272	166	.4897	794	7
54	299	200	.4876	784	6
55	.37326	.40234	2.485$\bar{5}$	.92773	5
56	353	267	.4834	762	4
57	380	301	.4813	751	3
58	407	33$\bar{5}$	.4792	740	2
59	434	369	.4772	729	1
60	.37461	.40403	2.4751	.92718	0
	cos	cot	tan	sin	′

68°

22° TABLE III 23°

22°

′	sin	tan	cot	cos	
0	.37461	.40403	2.4751	.92718	60
1	488	436	.4730	707	59
2	515	470	.4709	697	58
3	542	504	.4689	686	57
4	569	538	.4668	675	56
5	.37595	.40572	2.4648	.92664	55
6	622	606	.4627	653	54
7	649	640	.4606	642	53
8	676	674	.4586	631	52
9	703	707	.4566	620	51
10	.37730	.40741	2.4545	.92609	50
11	757	775	.4525	598	49
12	784	809	.4504	587	48
13	811	843	.4484	576	47
14	838	877	.4464	565	46
15	.37865	.40911	2.4443	.92554	45
16	892	945	.4423	543	44
17	919	.40979	.4403	532	43
18	946	.41013	.4383	521	42
19	973	047	.4362	510	41
20	.37999	.41081	2.4342	.92499	40
21	.38026	115	.4322	488	39
22	053	149	.4302	477	38
23	080	183	.4282	466	37
24	107	217	.4262	455	36
25	.38134	.41251	2.4242	.92444	35
26	161	285	.4222	432	34
27	188	319	.4202	421	33
28	215	353	.4182	410	32
29	241	387	.4162	399	31
30	.38268	.41421	2.4142	.92388	30
31	295	455	.4122	377	29
32	322	490	.4102	366	28
33	349	524	.4083	355	27
34	376	558	.4063	343	26
35	.38403	.41592	2.4043	.92332	25
36	430	626	.4023	321	24
37	456	660	.4004	310	23
38	483	694	.3984	299	22
39	510	728	.3964	287	21
40	.38537	.41763	2.3945	.92276	20
41	564	797	.3925	265	19
42	591	831	.3906	254	18
43	617	865	.3886	243	17
44	644	899	.3867	231	16
45	.38671	.41933	2.3847	.92220	15
46	698	.41968	.3828	209	14
47	725	.42002	.3808	198	13
48	752	036	.3789	186	12
49	778	070	.3770	175	11
50	.38805	.42105	2.3750	.92164	10
51	832	139	.3731	152	9
52	859	173	.3712	141	8
53	886	207	.3693	130	7
54	912	242	.3673	119	6
55	.38939	.42276	2.3654	.92107	5
56	966	310	.3635	096	4
57	.38993	345	.3616	085	3
58	.39020	379	.3597	073	2
59	046	413	.3578	062	1
60	.39073	.42447	2.3559	.92050	0
	cos	cot	tan	sin	′

67°

23°

′	sin	tan	cot	cos	
0	.39073	.42447	2.3559	.92050	60
1	100	482	.3539	039	59
2	127	516	.3520	028	58
3	153	551	.3501	016	57
4	180	585	.3483	.92005	56
5	.39207	.42619	2.3464	.91994	55
6	234	654	.3445	982	54
7	260	688	.3426	971	53
8	287	722	.3407	959	52
9	314	757	.3388	948	51
10	.39341	.42791	2.3369	.91936	50
11	367	826	.3351	925	49
12	394	860	.3332	914	48
13	421	894	.3313	902	47
14	448	929	.3294	891	46
15	.39474	.42963	2.3276	.91879	45
16	501	.42998	.3257	868	44
17	528	.43032	.3238	856	43
18	555	067	.3220	845	42
19	581	101	.3201	833	41
20	.39608	.43136	2.3183	.91822	40
21	635	170	.3164	810	39
22	661	205	.3146	799	38
23	688	239	.3127	787	37
24	715	274	.3109	775	36
25	.39741	.43308	2.3090	.91764	35
26	768	343	.3072	752	34
27	795	378	.3053	741	33
28	822	412	.3035	729	32
29	848	447	.3017	718	31
30	.39875	.43481	2.2998	.91706	30
31	902	516	.2980	694	29
32	928	550	.2962	683	28
33	955	585	.2944	671	27
34	.39982	620	.2925	660	26
35	.40008	.43654	2.2907	.91648	25
36	035	689	.2889	636	24
37	062	724	.2871	625	23
38	088	758	.2853	613	22
39	115	793	.2835	601	21
40	.40141	.43828	2.2817	.91590	20
41	168	862	.2799	578	19
42	195	897	.2781	566	18
43	221	932	.2763	555	17
44	248	.43966	.2745	543	16
45	.40275	.44001	2.2727	.91531	15
46	301	036	.2709	519	14
47	328	071	.2691	508	13
48	355	105	.2673	496	12
49	381	140	.2655	484	11
50	.40408	.44175	2.2637	.91472	10
51	434	210	.2620	461	9
52	461	244	.2602	449	8
53	488	279	.2584	437	7
54	514	314	.2566	425	6
55	.40541	.44349	2.2549	.91414	5
56	567	384	.2531	402	4
57	594	418	.2513	390	3
58	621	453	.2496	378	2
59	647	488	.2478	366	1
60	.40674	.44523	2.2460	.91355	0
	cos	cot	tan	sin	′

66°

24°

′	sin	tan	cot	cos	
0	.40674	.44523	2.2460	.91355	60
1	700	558	.2443	343	59
2	727	593	.2425	331	58
3	753	627	.2408	319	57
4	780	662	.2390	307	56
5	.40806	.44697	2.2373	.91295	55
6	833	732	.2355	283	54
7	860	767	.2338	272	53
8	886	802	.2320	260	52
9	913	837	.2303	248	51
10	.40939	.44872	2.2286	.91236	50
11	966	907	.2268	224	49
12	.40992	942	.2251	212	48
13	.41019	.44977	.2234	200	47
14	045	.45012	.2216	188	46
15	.41072	.45047	2.2199	.91176	45
16	098	082	.2182	164	44
17	125	117	.2165	152	43
18	151	152	.2148	140	42
19	178	187	.2130	128	41
20	.41204	.45222	2.2113	.91116	40
21	231	257	.2096	104	39
22	257	292	.2079	092	38
23	284	327	.2062	080	37
24	310	362	.2045	068	36
25	.41337	.45397	2.2028	.91056	35
26	363	432	.2011	044	34
27	390	467	.1994	032	33
28	416	502	.1977	020	32
29	443	538	.1960	.91008	31
30	.41469	.45573	2.1943	.90996	30
31	496	608	.1926	984	29
32	522	643	.1909	972	28
33	549	678	.1892	960	27
34	575	713	.1876	948	26
35	.41602	.45748	2.1859	.90936	25
36	628	784	.1842	924	24
37	655	819	.1825	911	23
38	681	854	.1808	899	22
39	707	889	.1792	887	21
40	.41734	.45924	2.1775	.90875	20
41	760	960	.1758	863	19
42	787	.45995	.1742	851	18
43	813	.46030	.1725	839	17
44	840	065	.1708	826	16
45	.41866	.46101	2.1692	.90814	15
46	892	136	.1675	802	14
47	919	171	.1659	790	13
48	945	206	.1642	778	12
49	972	242	.1625	766	11
50	.41998	.46277	2.1609	.90753	10
51	.42024	312	.1592	741	9
52	051	348	.1576	729	8
53	077	383	.1560	717	7
54	104	418	.1543	704	6
55	.42130	.46454	2.1527	.90692	5
56	156	489	.1510	680	4
57	183	525	.1494	668	3
58	209	560	.1478	655	2
59	235	595	.1461	643	1
60	.42262	.46631	2.1445	.90631	0
	cos	cot	tan	sin	′

65°

25°

′	sin	tan	cot	cos	
0	.42262	.46631	2.1445	.90631	60
1	288	666	.1429	618	59
2	315	702	.1413	606	58
3	341	737	.1396	594	57
4	367	772	.1380	582	56
5	.42394	.46808	2.1364	.90569	55
6	420	843	.1348	557	54
7	446	879	.1332	545	53
8	473	914	.1315	532	52
9	499	950	.1299	520	51
10	.42525	.46985	2.1283	.90507	50
11	552	.47021	.1267	495	49
12	578	056	.1251	483	48
13	604	092	.1235	470	47
14	631	128	.1219	458	46
15	.42657	.47163	2.1203	.90446	45
16	683	199	.1187	433	44
17	709	234	.1171	421	43
18	736	270	.1155	408	42
19	762	305	.1139	396	41
20	.42788	.47341	2.1123	.90383	40
21	815	377	.1107	371	39
22	841	412	.1092	358	38
23	867	448	.1076	346	37
24	894	483	.1060	334	36
25	.42920	.47519	2.1044	.90321	35
26	946	555	.1028	309	34
27	972	590	.1013	296	33
28	.42999	626	0997	284	32
29	.43025	662	.0981	271	31
30	.43051	.47698	2.0965	.90259	30
31	077	733	.0950	246	29
32	104	769	.0934	233	28
33	130	805	.0918	221	27
34	156	840	.0903	208	26
35	.43182	.47876	2.0887	.90196	25
36	209	912	.0872	183	24
37	235	948	.0856	171	23
38	261	.47984	.0840	158	22
39	287	.48019	.0825	146	21
40	.43313	.48055	2.0809	.90133	20
41	340	091	.0794	120	19
42	366	127	.0778	108	18
43	392	163	.0763	095	17
44	418	198	.0748	082	16
45	.43445	.48234	2.0732	.90070	15
46	471	270	.0717	057	14
47	497	306	.0701	045	13
48	523	342	.0686	032	12
49	549	378	.0671	019	11
50	.43575	.48414	2.0655	.90007	10
51	602	450	.0640	.89994	9
52	628	486	.0625	981	8
53	654	521	.0609	968	7
54	680	557	.0594	956	6
55	.43706	.48593	2.0579	.89943	5
56	733	629	.0564	930	4
57	759	665	.0549	918	3
58	785	701	.0533	905	2
59	811	737	.0518	892	1
60	.43837	.48773	2.0503	.89879	0
	cos	cot	tan	sin	′

64°

26°

′	sin	tan	cot	cos	
0	.43837	.48773	2.0503	.89879	60
1	863	809	.0488	867	59
2	889	845	.0473	854	58
3	916	881	.0458	841	57
4	942	917	.0443	828	56
5	.43968	.48953	2.0428	.89816	55
6	.43994	.48989	.0413	803	54
7	.44020	.49026	.0398	790	53
8	046	062	.0383	777	52
9	072	098	.0368	764	51
10	.44098	.49134	2.0353	.89752	50
11	124	170	.0338	739	49
12	151	206	.0323	726	48
13	177	242	.0308	713	47
14	203	278	.0293	700	46
15	.44229	.49315	2.0278	.89687	45
16	255	351	.0263	674	44
17	281	387	.0248	662	43
18	307	423	.0233	649	42
19	333	459	.0219	636	41
20	.44359	.49495	2.0204	.89623	40
21	385	532	.0189	610	39
22	411	568	.0174	597	38
23	437	604	.0160	584	37
24	464	640	.0145	571	36
25	.44490	.49677	2.0130	.89558	35
26	516	713	.0115	545	34
27	542	749	.0101	532	33
28	568	786	.0086	519	32
29	594	822	.0072	506	31
30	.44620	.49858	2.0057	.89493	30
31	646	894	.0042	480	29
32	672	931	.0028	467	28
33	698	.49967	2.0013	454	27
34	724	.50004	1.9999	441	26
35	.44750	.50040	1.9984	.89428	25
36	776	076	.9970	415	24
37	802	113	.9955	402	23
38	828	149	.9941	389	22
39	854	185	.9926	376	21
40	.44880	.50222	1.9912	.89363	20
41	906	258	.9897	350	19
42	932	295	.9883	337	18
43	958	331	.9868	324	17
44	.44984	368	.9854	311	16
45	.45010	.50404	1.9840	.89298	15
46	036	441	.9825	285	14
47	062	477	.9811	272	13
48	088	514	.9797	259	12
49	114	550	.9782	245	11
50	.45140	.50587	1.9768	.89232	10
51	166	623	.9754	219	9
52	192	660	.9740	206	8
53	218	696	9725	193	7
54	243	733	.9711	180	6
55	.45269	.50769	1.9697	.89167	5
56	295	806	.9683	153	4
57	321	843	.9669	140	3
58	347	879	.9654	127	2
59	373	916	.9640	114	1
60	.45399	.50953	1.9626	.89101	0
	cos	cot	tan	sin	′

63°

27°

′	sin	tan	cot	cos	
0	.45399	.50953	1.9626	.89101	60
1	425	.50989	.9612	087	59
2	451	.51026	.9598	074	58
3	477	063	.9584	061	57
4	503	.099	.9570	048	56
5	.45529	.51136	1.9556	.89035	55
6	554	173	.9542	021	54
7	580	209	.9528	.89008	53
8	606	246	.9514	.88995	52
9	632	283	.9500	981	51
10	.45658	.51319	1.9486	.88968	50
11	684	356	.9472	955	49
12	710	393	.9458	942	48
13	736	430	.9444	928	47
14	762	467	.9430	915	46
15	.45787	.51503	1.9416	.88902	45
16	813	540	.9402	888	44
17	839	577	.9388	875	43
18	865	614	.9375	862	42
19	891	651	.9361	848	41
20	.45917	.51688	1.9347	.88835	40
21	942	724	.9333	822	39
22	968	761	.9319	808	38
23	.45994	798	.9306	795	37
24	.46020	835	.9292	782	36
25	.46046	.51872	1.9278	.88768	35
26	072	909	.9265	755	34
27	097	946	.9251	741	33
28	123	.51983	.9237	728	32
29	149	.52020	.9223	715	31
30	.46175	.52057	1.9210	.88701	30
31	201	094	.9196	688	29
32	226	131	.9183	674	28
33	252	168	.9169	661	27
34	278	205	.9155	647	26
35	.46304	.52242	1.9142	.88634	25
36	330	279	.9128	620	24
37	355	316	.9115	607	23
38	381	353	.9101	593	22
39	407	390	.9088	580	21
40	.46433	.52427	1.9074	.88566	20
41	458	464	.9061	553	19
42	484	501	.9047	539	18
43	510	538	.9034	526	17
44	536	575	.9020	512	16
45	.46561	.52613	1.9007	.88499	15
46	587	650	.8993	485	14
47	613	687	.8980	472	13
48	639	724	.8967	458	12
49	664	761	.8953	445	11
50	.46690	.52798	1.8940	.88431	10
51	716	836	.8927	417	9
52	742	873	.8913	404	8
53	767	910	.8900	390	7
54	793	947	.8887	377	6
55	.46819	.52985	1.8873	.88363	5
56	844	.53022	.8860	349	4
57	870	059	.8847	336	3
58	896	096	.8834	322	2
59	921	134	.8820	308	1
60	.46947	.53171	1.8807	.88295	0
	cos	cot	tan	sin	′

62°

28°

′	sin	tan	cot	cos	
0	.46947	.53171	1.8807	.88295	**60**
1	973	208	.8794	281	59
2	.46999	246	.8781	267	58
3	.47024	283	.8768	254	57
4	050	320	.8755	240	56
5	.47076	.53358	1.8741	.88226	**55**
6	101	395	.8728	213	54
7	127	432	.8715	199	53
8	153	470	.8702	185	52
9	178	507	.8689	172	51
10	.47204	.53545	1.8676	.88158	**50**
11	229	582	.8663	144	49
12	255	620	.8650	130	48
13	281	657	.8637	117	47
14	306	694	.8624	103	46
15	.47332	.53732	1.8611	.88089	**45**
16	358	769	.8598	075	44
17	383	807	.8585	062	43
18	409	844	.8572	048	42
19	434	882	.8559	034	41
20	.47460	.53920	1.8546	.88020	**40**
21	486	957	.8533	.88006	39
22	511	.53995	.8520	.87993	38
23	537	.54032	.8507	979	37
24	562	070	.8495	965	36
25	.47588	.54107	1.8482	.87951	**35**
26	614	145	.8469	937	34
27	639	183	.8456	923	33
28	665	220	.8443	909	32
29	690	258	.8430	896	31
30	.47716	.54296	1.8418	.87882	**30**
31	741	333	.8405	868	29
32	767	371	.8392	854	28
33	793	409	.8379	840	27
34	818	446	.8367	826	26
35	.47844	.54484	1.8354	.87812	**25**
36	869	522	.8341	798	24
37	895	560	.8329	784	23
38	920	597	.8316	770	22
39	946	635	.8303	756	21
40	.47971	.54673	1.8291	.87743	**20**
41	.47997	711	.8278	729	19
42	.48022	748	.8265	715	18
43	048	786	.8253	701	17
44	073	824	.8240	687	16
45	.48099	.54862	1.8228	.87673	**15**
46	124	900	.8215	659	14
47	150	938	.8202	645	13
48	175	.54975	.8190	631	12
49	201	.55013	.8177	617	11
50	.48226	.55051	1.8165	.87603	**10**
51	252	089	.8152	589	9
52	277	127	.8140	575	8
53	303	165	.8127	561	7
54	328	203	.8115	546	6
55	.48354	.55241	1.8103	.87532	**5**
56	379	279	.8090	518	4
57	405	317	.8078	504	3
58	430	355	.8065	490	2
59	456	393	.8053	476	1
60	.48481	.55431	1.8040	.87462	**0**
	cos	cot	tan	sin	′

61°

29°

′	sin	tan	cot	cos	
0	.48481	.55431	1.8040	.87462	**60**
1	506	469	.8028	448	59
2	532	507	.8016	434	58
3	557	545	.8003	420	57
4	583	583	.7991	406	56
5	.48608	.55621	1.7979	.87391	**55**
6	634	659	.7966	377	54
7	659	697	.7954	363	53
8	684	736	.7942	349	52
9	710	774	.7930	335	51
10	.48735	.55812	1.7917	.87321	**50**
11	761	850	.7905	306	49
12	786	888	.7893	292	48
13	811	926	.7881	278	47
14	837	.55964	.7868	264	46
15	.48862	.56003	1.7856	.87250	**45**
16	888	041	.7844	235	44
17	913	079	.7832	221	43
18	938	117	.7820	207	42
19	964	156	.7808	193	41
20	.48989	.56194	1.7796	.87178	**40**
21	.49014	232	.7783	164	39
22	040	270	.7771	150	38
23	065	309	.7759	136	37
24	090	347	.7747	121	36
25	.49116	.56385	1.7735	.87107	**35**
26	141	424	.7723	093	34
27	166	462	.7711	079	33
28	192	501	.7699	064	32
29	217	539	.7687	050	31
30	.49242	.56577	1.7675	.87036	**30**
31	268	616	.7663	021	29
32	293	654	.7651	.87007	28
33	318	693	.7639	.86993	27
34	344	731	7627	978	26
35	.49369	.56769	1.7615	.86964	**25**
36	394	808	.7603	949	24
37	419	846	.7591	935	23
38	445	885	.7579	921	22
39	470	923	.7567	906	21
40	.49495	.56962	1.7556	.86892	**20**
41	521	.57000	.7544	878	19
42	546	039	.7532	863	18
43	571	078	.7520	849	17
44	596	116	.7508	834	16
45	.49622	.57155	1.7496	.86820	**15**
46	647	193	.7485	805	14
47	672	232	.7473	791	13
48	697	271	.7461	777	12
49	723	309	.7449	762	11
50	.49748	.57348	1.7437	.86748	**10**
51	773	386	.7426	733	9
52	798	425	.7414	719	8
53	824	464	.7402	704	7
54	849	503	.7391	690	6
55	.49874	.57541	1.7379	.86675	**5**
56	899	580	.7367	661	4
57	924	619	.7355	646	3
58	950	657	.7344	632	2
59	.49975	696	.7332	617	1
60	.50000	.57735	1.7321	.86603	**0**
	cos	cot	tan	sin	′

60°

30° TABLE III

′	sin	tan	cot	cos	′
0	.50000	.57735	1.7321	.86603	60
1	025	774	.7309	588	59
2	050	813	.7297	573	58
3	076	851	.7286	559	57
4	101	890	.7274	544	56
5	.50126	.57929	1.7262	.86530	55
6	151	.57968	.7251	515	54
7	176	.58007	.7239	501	53
8	201	046	.7228	486	52
9	227	08$\bar{5}$	.7216	471	51
10	.50252	.58124	1.720$\bar{5}$	.86457	50
11	277	162	.7193	442	49
12	302	201	.7182	427	48
13	327	240	.7170	413	47
14	352	279	.7159	398	46
15	.50377	.58318	1.7147	.86384	45
16	403	357	.7136	369	44
17	428	396	.7124	354	43
18	453	435	.7113	340	42
19	478	474	.7102	32$\bar{5}$	41
20	.50503	.58513	1.7090	.86310	40
21	528	552	.7079	295	39
22	553	591	.7067	281	38
23	578	631	.7056	266	37
24	603	670	.704$\bar{5}$	251	36
25	.50628	.58709	1.7033	.86237	35
26	654	748	.7022	222	34
27	679	787	.7011	207	33
28	704	826	.6999	192	32
29	729	865	.6988	178	31
30	.50754	.5890$\bar{5}$	1.6977	.86163	30
31	779	944	.6965	148	29
32	804	.58983	.6954	133	28
33	829	.59022	.6943	119	27
34	854	061	.6932	104	26
35	.50879	.59101	1.6920	.86089	25
36	904	140	.6909	074	24
37	929	179	.6898	059	23
38	954	218	.6887	04$\bar{5}$	22
39	.50979	258	.6875	030	21
40	.51004	.59297	1.6864	.8601$\bar{5}$	20
41	029	336	.6853	.86000	19
42	054	376	.6842	.85985	18
43	079	415	.6831	970	17
44	104	454	.6820	956	16
45	.51129	.59494	1.6808	.85941	15
46	154	533	.6797	926	14
47	179	573	.6786	911	13
48	204	612	.6775	896	12
49	229	651	.6764	881	11
50	.51254	.59691	1.6753	.85866	10
51	279	730	.6742	851	9
52	304	770	.6731	836	8
53	329	809	.6720	821	7
54	354	849	.6709	806	6
55	.51379	.59888	1.6698	.85792	5
56	404	928	.6687	777	4
57	429	.59967	.6676	762	3
58	454	.60007	.666$\bar{5}$	747	2
59	479	046	.6654	732	1
60	.51504	.60086	1.6643	.85717	0
′	cos	cot	tan	sin	′

59°

31°

′	sin	tan	cot	cos	′
0	.51504	.60086	1.6643	.85717	60
1	529	126	.6632	702	59
2	554	165	.6621	687	58
3	579	20$\bar{5}$	.6610	672	57
4	604	24$\bar{5}$	.6599	657	56
5	.51628	.60284	1.6588	.85642	55
6	653	324	.6577	627	54
7	678	364	.6566	612	53
8	703	403	.6555	597	52
9	728	443	.654$\bar{5}$	582	51
10	.51753	.60483	1.6534	.85567	50
11	778	522	.6523	551	49
12	803	562	.6512	536	48
13	828	602	.6501	521	47
14	852	642	.6490	506	46
15	.51877	.60681	1.6479	.85491	45
16	902	721	.6469	476	44
17	927	761	.6458	461	43
18	952	801	.6447	446	42
19	.51977	841	.6436	431	41
20	.52002	.60881	1.6426	.85416	40
21	026	921	.641$\bar{5}$	401	39
22	051	.60960	.6404	385	38
23	076	.61000	.6393	370	37
24	101	040	.6383	355	36
25	.52126	.61080	1.6372	.85340	35
26	151	120	.6361	32$\bar{5}$	34
27	175	160	.6351	310	33
28	200	200	.6340	294	32
29	225	240	.6329	279	31
30	.522$\bar{5}$0	.61280	1.6319	.85264	30
31	27$\bar{5}$	320	.6308	249	29
32	299	360	.6297	234	28
33	324	400	.6287	218	27
34	349	440	.6276	203	26
35	.52374	.61480	1.6265	.85188	25
36	399	520	.625$\bar{5}$	173	24
37	423	561	.6244	157	23
38	448	601	.6234	142	22
39	473	641	.6223	127	21
40	.52498	.61681	1.6212	.85112	20
41	522	721	.6202	096	19
42	547	761	.6191	081	18
43	572	801	.6181	066	17
44	597	842	.6170	051	16
45	.52621	.61882	1.6160	.85035	15
46	646	922	.6149	020	14
47	671	.61962	.6139	.8500$\bar{5}$	13
48	696	.62003	.6128	.84989	12
49	720	043	.6118	974	11
50	.52745	.62083	1.6107	.84959	10
51	770	124	.6097	943	9
52	794	164	.6087	928	8
53	819	204	.6076	913	7
54	844	24$\bar{5}$	.6066	897	6
55	.52869	.6228$\bar{5}$	1.6055	.84882	5
56	893	325	.604$\bar{5}$	866	4
57	918	366	.6034	851	3
58	943	406	.6024	836	2
59	967	446	.6014	820	1
60	.52992	.62487	1.6003	.8480$\bar{5}$	0
′	cos	cot	tan	sin	′

58°

32°

′	sin	tan	cot	cos	
0	.52992	.62487	1.6003	.84805	60
1	.53017	527	.5993	789	59
2	041	568	.5983	774	58
3	066	608	.5972	759	57
4	091	649	.5962	743	56
5	.53115	.62689	1.5952	.84728	55
6	140	730	.5941	712	54
7	164	770	.5931	697	53
8	189	811	.5921	681	52
9	214	852	.5911	666	51
10	.53238	.62892	1.5900	.84650	50
11	263	933	.5890	635	49
12	288	.62973	.5880	619	48
13	312	.63014	.5869	604	47
14	337	055	.5859	588	46
15	.53361	.63095	1.5849	.84573	45
16	386	136	.5839	557	44
17	411	177	.5829	542	43
18	435	217	.5818	526	42
19	460	258	.5808	511	41
20	.53484	.63299	1.5798	.84495	40
21	509	340	.5788	480	39
22	534	380	.5778	464	38
23	558	421	.5768	448	37
24	583	462	.5757	433	36
25	.53607	.63503	1.5747	.84417	35
26	632	544	.5737	402	34
27	656	584	.5727	386	33
28	681	625	.5717	370	32
29	705	666	.5707	355	31
30	.53730	.63707	1.5697	.84339	30
31	754	748	.5687	324	29
32	779	789	.5677	308	28
33	804	830	.5667	292	27
34	828	871	.5657	277	26
35	.53853	.63912	1.5647	.84261	25
36	877	953	.5637	245	24
37	902	.63994	.5627	230	23
38	926	.64035	.5617	214	22
39	951	076	.5607	198	21
40	.53975	.64117	1.5597	.84182	20
41	.54000	158	.5587	167	19
42	024	199	.5577	151	18
43	049	240	.5567	135	17
44	073	281	.5557	120	16
45	.54097	.64322	1.5547	.84104	15
46	122	363	.5537	088	14
47	146	404	.5527	072	13
48	171	446	.5517	057	12
49	195	487	.5507	041	11
50	.54220	.64528	1.5497	.84025	10
51	244	569	.5487	.84009	9
52	269	610	.5477	.83994	8
53	293	652	.5468	978	7
54	317	693	.5458	962	6
55	.54342	.64734	1.5448	.83946	5
56	366	775	.5438	930	4
57	391	817	.5428	915	3
58	415	858	.5418	899	2
59	440	899	.5408	883	1
60	.54464	.64941	1.5399	.83867	0
	cos	cot	tan	sin	′

57°

33°

′	sin	tan	cot	cos	
0	.54464	.64941	1.5399	.83867	60
1	488	.64982	.5389	851	59
2	513	.65024	.5379	835	58
3	537	065	.5369	819	57
4	561	106	.5359	804	56
5	.54586	.65148	1.5350	.83788	55
6	610	189	.5340	772	54
7	635	231	.5330	756	53
8	659	272	.5320	740	52
9	683	314	.5311	724	51
10	.54708	.65355	1.5301	.83708	50
11	732	397	.5291	692	49
12	756	438	.5282	676	48
13	781	480	.5272	660	47
14	805	521	.5262	645	46
15	.54829	.65563	1.5253	.83629	45
16	854	604	.5243	613	44
17	878	646	.5233	597	43
18	902	688	.5224	581	42
19	927	729	.5214	565	41
20	.54951	.65771	1.5204	.83549	40
21	975	813	.5195	533	39
22	.54999	854	.5185	517	38
23	.55024	896	.5175	501	37
24	048	938	.5166	485	36
25	.55072	.65980	1.5156	.83469	35
26	097	.66021	.5147	453	34
27	121	063	.5137	437	33
28	145	105	.5127	421	32
29	169	147	.5118	405	31
30	.55194	.66189	1.5108	.83389	30
31	218	230	.5099	373	29
32	242	272	.5089	356	28
33	266	314	.5080	340	27
34	291	356	.5070	324	26
35	.55315	.66398	1.5061	.83308	25
36	339	440	.5051	292	24
37	363	482	.5042	276	23
38	388	524	.5032	260	22
39	412	566	.5023	244	21
40	.55436	.66608	1.5013	.83228	20
41	460	650	.5004	212	19
42	484	692	.4994	195	18
43	509	734	.4985	179	17
44	533	776	.4975	163	16
45	.55557	.66818	1.4966	.83147	15
46	581	860	.4957	131	14
47	605	902	.4947	115	13
48	630	944	.4938	098	12
49	654	.66986	.4928	082	11
50	.55678	.67028	1.4919	.83066	10
51	702	071	.4910	050	9
52	726	113	.4900	034	8
53	750	155	.4891	017	7
54	775	197	.4882	.83001	6
55	.55799	.67239	1.4872	.82985	5
56	823	282	.4863	969	4
57	847	324	.4854	953	3
58	871	366	.4844	936	2
59	895	409	.4835	920	1
60	.55919	.67451	1.4826	.82904	0
	cos	cot	tan	sin	′

56°

34°

′	sin	tan	cot	cos	
0	.55919	.67451	1.4826	.82904	60
1	943	493	.4816	887	59
2	968	536	.4807	871	58
3	.55992	578	.4798	855	57
4	.56016	620	.4788	839	56
5	.56040	.67663	1.4779	.82822	55
6	064	705	.4770	806	54
7	088	748	.4761	790	53
8	112	790	.4751	773	52
9	136	832	.4742	757	51
10	.56160	.67875	1.4733	.82741	50
11	184	917	.4724	724	49
12	208	.67960	.4715	708	48
13	232	.68002	.4705	692	47
14	256	045	.4696	675	46
15	.56280	.68088	1.4687	.82659	45
16	305	130	.4678	643	44
17	329	173	.4669	626	43
18	353	215	.4659	610	42
19	377	258	.4650	593	41
20	.56401	.68301	1.4641	.82577	40
21	425	343	.4632	561	39
22	449	386	.4623	544	38
23	473	429	.4614	528	37
24	497	471	.4605	511	36
25	.56521	.68514	1.4596	.82495	35
26	545	557	.4586	478	34
27	569	600	.4577	462	33
28	593	642	.4568	446	32
29	617	685	.4559	429	31
30	.56641	.68728	1.4550	.82413	30
31	665	771	.4541	396	29
32	689	814	.4532	380	28
33	713	857	.4523	363	27
34	736	900	.4514	347	26
35	.56760	.68942	1.4505	.82330	25
36	784	.68985	.4496	314	24
37	808	.69028	.4487	297	23
38	832	071	.4478	281	22
39	856	114	.4469	264	21
40	.56880	.69157	1.4460	.82248	20
41	904	200	.4451	231	19
42	928	243	.4442	214	18
43	952	286	.4433	198	17
44	.56976	329	.4424	181	16
45	.57000	.69372	1.4415	.82165	15
46	024	416	.4406	148	14
47	047	459	.4397	132	13
48	071	502	.4388	115	12
49	095	545	.4379	098	11
50	.57119	.69588	1.4370	.82082	10
51	143	631	.4361	065	9
52	167	675	.4352	048	8
53	191	718	.4344	032	7
54	215	761	.4335	.82015	6
55	.57238	.69804	1.4326	.81999	5
56	262	847	.4317	982	4
57	286	891	.4308	965	3
58	310	934	.4299	949	2
59	334	.69977	.4290	932	1
60	.57358	.70021	1.4281	.81915	0
	cos	cot	tan	sin	′

55°

35°

′	sin	tan	cot	cos	
0	.57358	.70021	1.4281	.81915	60
1	381	064	.4273	899	59
2	405	107	.4264	882	58
3	429	151	.4255	865	57
4	453	194	.4246	848	56
5	.57477	.70238	1.4237	.81832	55
6	501	281	.4229	815	54
7	524	325	.4220	798	53
8	548	368	.4211	782	52
9	572	412	.4202	765	51
10	.57596	.70455	1.4193	.81748	50
11	619	499	.4185	731	49
12	643	542	.4176	714	48
13	667	586	.4167	698	47
14	691	629	.4158	681	46
15	.57715	.70673	1.4150	.81664	45
16	738	717	.4141	647	44
17	762	760	.4132	631	43
18	786	804	.4124	614	42
19	810	848	.4115	597	41
20	.57833	.70891	1.4106	.81580	40
21	857	935	.4097	563	39
22	881	.70979	.4089	546	38
23	904	.71023	.4080	530	37
24	928	066	.4071	513	36
25	.57952	.71110	1.4063	.81496	35
26	976	154	.4054	479	34
27	.57999	198	.4045	462	33
28	.58023	242	.4037	445	32
29	047	285	.4028	428	31
30	.58070	.71329	1.4019	.81412	30
31	094	373	.4011	395	29
32	118	417	.4002	378	28
33	141	461	.3994	361	27
34	165	505	.3985	344	26
35	.58189	.71549	1.3976	.81327	25
36	212	593	.3968	310	24
37	236	637	.3959	293	23
38	260	681	.3951	276	22
39	283	725	.3942	259	21
40	.58307	.71769	1.3934	.81242	20
41	330	813	.3925	225	19
42	354	857	.3916	208	18
43	378	901	.3908	191	17
44	401	946	.3899	174	16
45	.58425	.71990	1.3891	.81157	15
46	449	.72034	.3882	140	14
47	472	078	.3874	123	13
48	496	122	.3865	106	12
49	519	167	.3857	089	11
50	.58543	.72211	1.3848	.81072	10
51	567	255	.3840	055	9
52	590	299	.3831	038	8
53	614	344	.3823	021	7
54	637	388	.3814	.81004	6
55	.58661	.72432	1.3806	80987	5
56	684	477	.3798	970	4
57	708	521	.3789	953	3
58	731	565	.3781	936	2
59	755	610	.3772	919	1
60	.58779	.72654	1.3764	.80902	0
	cos	cot	tan	sin	′

54°

36°

′	sin	tan	cot	cos	
0	.58779	.72654	1.3764	.80902	**60**
1	802	699	.3755	88$\bar{5}$	59
2	826	743	.3747	867	58
3	849	788	.3739	850	57
4	873	832	.3730	833	56
5	.58896	.72877	1.3722	.80816	**55**
6	920	921	.3713	799	54
7	943	.72966	.3705	782	53
8	967	.73010	.3697	76$\bar{5}$	52
9	.58990	055	.3688	748	51
10	.59014	.73100	1.3680	.80730	**50**
11	037	144	.3672	713	49
12	061	189	.3663	696	48
13	084	234	.365$\bar{5}$	679	47
14	108	278	.3647	662	46
15	.59131	.73323	1.3638	.80644	**45**
16	154	368	.3630	627	44
17	178	413	.3622	610	43
18	201	457	.3613	593	42
19	22$\bar{5}$	502	.3605	576	41
20	.59248	.73547	1.3597	.80558	**40**
21	272	592	.3588	541	39
22	295	637	.3580	524	38
23	318	681	.3572	507	37
24	342	726	.3564	489	36
25	.59365	.73771	1.3555	.80472	**35**
26	389	816	.3547	45$\bar{5}$	34
27	412	861	.3539	438	33
28	436	906	.3531	420	32
29	459	951	.3522	403	31
30	.59482	.73996	1.3514	.80386	**30**
31	506	74041	.3506	368	29
32	529	086	.3498	351	28
33	552	131	.3490	334	27
34	576	176	.3481	316	26
35	.59599	.74221	1.3473	.80299	**25**
36	622	267	.3465	282	24
37	646	312	.3457	264	23
38	669	357	.3449	247	22
39	693	402	.3440	230	21
40	.59716	.74447	1.3432	.80212	**20**
41	739	492	.3424	19$\bar{5}$	19
42	763	538	.3416	178	18
43	786	583	.3408	160	17
44	809	628	.3400	143	16
45	.59832	.74674	1.3392	.80125	**15**
46	856	719	.3384	108	14
47	879	764	.3375	091	13
48	902	810	.3367	073	12
49	926	85$\bar{5}$	.3359	056	11
50	.59949	.74900	1.3351	.80038	**10**
51	972	946	.3343	021	9
52	.59995	.74991	.333$\bar{5}$	.80003	8
53	.60019	.75037	.3327	.79986	7
54	042	082	.3319	968	6
55	.60065	.75128	1.3311	.79951	**5**
56	089	173	.3303	934	4
57	112	219	.329$\bar{5}$	916	3
58	135	264	.3287	899	2
59	158	310	.3278	881	1
60	.60182	.75355	1.3270	.79864	**0**
	cos	cot	tan	sin	′

53°

37°

′	sin	tan	cot	cos	
0	.60182	.75355	1.3270	.79864	**60**
1	20$\bar{5}$	401	.3262	846	59
2	228	447	.3254	829	58
3	251	492	.3246	811	57
4	274	538	.3238	793	56
5	.60298	.75584	1.3230	.79776	**55**
6	321	629	.3222	758	54
7	344	675	.3214	741	53
8	367	721	.3206	723	52
9	390	767	.3198	706	51
10	.60414	.75812	1.3190	.79688	**50**
11	437	858	.3182	671	49
12	460	904	.3175	653	48
13	483	9$\bar{5}$0	.3167	635	47
14	506	.75996	.3159	618	46
15	.60529	.76042	1.3151	.79600	**45**
16	553	088	.3143	583	44
17	576	134	.3135	56$\bar{5}$	43
18	599	180	.3127	547	42
19	622	226	.3119	530	41
20	.60645	.76272	1.3111	.79512	**40**
21	668	318	.3103	494	39
22	691	364	.3095	477	38
23	714	410	.3087	459	37
24	738	456	.3079	441	36
25	.60761	.76502	1.3072	.79424	**35**
26	784	548	.3064	406	34
27	807	594	.3056	388	33
28	830	640	.3048	371	32
29	853	686	.3040	353	31
30	.60876	.76733	1.3032	.79335	**30**
31	899	779	.3024	318	29
32	922	825	.3017	300	28
33	945	871	.3009	282	27
34	968	918	.3001	264	26
35	.60991	.76964	1.2993	.79247	**25**
36	.6101$\bar{5}$	.77010	.2985	229	24
37	038	057	.2977	211	23
38	061	103	.2970	193	22
39	084	149	.2962	176	21
40	.61107	.77196	1.2954	.79158	**20**
41	130	242	.2946	140	19
42	153	289	.2938	122	18
43	176	335	.2931	10$\bar{5}$	17
44	199	382	.2923	087	16
45	.61222	.77428	1.2915	.79069	**15**
46	24$\bar{5}$	47$\bar{5}$	.2907	051	14
47	268	521	.2900	033	13
48	291	568	.2892	.79016	12
49	314	61$\bar{5}$	.2884	.78998	11
50	.61337	.77661	1.2876	.78980	**10**
51	360	708	.2869	962	9
52	383	754	.2861	944	8
53	406	801	.2853	926	7
54	429	848	.2846	908	6
55	.61451	.7789$\bar{5}$	1.2838	.78891	**5**
56	474	941	.2830	873	4
57	497	.77988	.2822	85$\bar{5}$	3
58	520	.7803$\bar{5}$	.2815	837	2
59	543	082	.2807	819	1
60	.61566	.78129	1.2799	.78801	**0**
	cos	cot	tan	sin	′

52°

′	sin	tan	cot	cos	
0	.61566	.78129	1.2799	.78801	60
1	589	175	.2792	783	59
2	612	222	.2784	765	58
3	635	269	.2776	747	57
4	658	316	.2769	729	56
5	.61681	.78363	1.2761	.78711	55
6	704	410	.2753	694	54
7	726	457	.2746	676	53
8	749	504	.2738	658	52
9	772	551	.2731	640	51
10	.61795	.78598	1.2723	.78622	50
11	818	645	.2715	604	49
12	841	692	.2708	586	48
13	864	739	.2700	568	47
14	887	786	.2693	550	46
15	.61909	.78834	1.2685	.78532	45
16	932	881	.2677	514	44
17	955	928	.2670	496	43
18	.61978	.78975	.2662	478	42
19	.62001	.79022	.2655	460	41
20	.62024	.79070	1.2647	.78442	40
21	046	117	.2640	424	39
22	069	164	.2632	405	38
23	092	212	.2624	387	37
24	115	259	.2617	369	36
25	.62138	.79306	1.2609	.78351	35
26	160	354	.2602	333	34
27	183	401	.2594	315	33
28	206	449	.2587	297	32
29	229	496	.2579	279	31
30	.62251	.79544	1.2572	.78261	30
31	274	591	.2564	243	29
32	297	639	.2557	225	28
33	320	686	.2549	206	27
34	342	734	.2542	188	26
35	.62365	.79781	1.2534	.78170	25
36	388	829	.2527	152	24
37	411	877	.2519	134	23
38	433	924	.2512	116	22
39	456	.79972	.2504	098	21
40	.62479	.80020	1.2497	.78079	20
41	502	067	.2489	061	19
42	524	115	.2482	043	18
43	547	163	.2475	025	17
44	570	211	.2467	.78007	16
45	.62592	.80258	1.2460	.77988	15
46	615	306	.2452	970	14
47	638	354	.2445	952	13
48	660	402	.2437	934	12
49	683	450	.2430	916	11
50	.62706	.80498	1.2423	.77897	10
51	728	546	.2415	879	9
52	751	594	.2408	861	8
53	774	642	.2401	843	7
54	796	690	.2393	824	6
55	.62819	.80738	1.2386	.77806	5
56	842	786	.2378	788	4
57	864	834	.2371	769	3
58	887	882	.2364	751	2
59	909	930	.2356	733	1
60	.62932	.80978	1.2349	.77715	0
	cos	cot	tan	sin	′

′	sin	tan	cot	cos	
0	.62932	.80978	1.2349	.77715	60
1	955	.81027	.2342	696	59
2	.62977	075	.2334	678	58
3	.63000	123	.2327	660	57
4	022	171	.2320	641	56
5	.63045	.81220	1.2312	.77623	55
6	068	268	.2305	605	54
7	090	316	.2298	586	53
8	113	364	.2290	568	52
9	135	413	.2283	550	51
10	.63158	.81461	1.2276	.77531	50
11	180	510	.2268	513	49
12	203	558	.2261	494	48
13	225	606	.2254	476	47
14	248	655	.2247	458	46
15	.63271	.81703	1.2239	.77439	45
16	293	752	.2232	421	44
17	316	800	.2225	402	43
18	338	849	.2218	384	42
19	361	898	.2210	366	41
20	.63383	.81946	1.2203	.77347	40
21	406	.81995	.2196	329	39
22	428	.82044	.2189	310	38
23	451	092	.2181	292	37
24	473	141	.2174	273	36
25	.63496	.82190	1.2167	.77255	35
26	518	238	.2160	236	34
27	540	287	.2153	218	33
28	563	336	.2145	199	32
29	585	385	.2138	181	31
30	.63608	.82434	1.2131	.77162	30
31	630	483	.2124	144	29
32	653	531	.2117	125	28
33	675	580	.2109	107	27
34	698	629	.2102	088	26
35	.63720	.82678	1.2095	.77070	25
36	742	727	.2088	051	24
37	765	776	.2081	033	23
38	787	825	.2074	.77014	22
39	810	874	.2066	.76996	21
40	.63832	.82923	1.2059	.76977	20
41	854	.82972	.2052	959	19
42	877	.83022	.2045	940	18
43	899	071	.2038	921	17
44	922	120	.2031	903	16
45	.63944	.83169	1.2024	.76884	15
46	966	218	.2017	866	14
47	.63989	268	.2009	847	13
48	.64011	317	.2002	828	12
49	033	366	.1995	810	11
50	.64056	.83415	1.1988	.76791	10
51	078	465	.1981	772	9
52	100	514	.1974	754	8
53	123	564	.1967	735	7
54	145	613	.1960	717	6
55	.64167	.83662	1.1953	.76698	5
56	190	712	.1946	679	4
57	212	761	.1939	661	3
58	234	811	.1932	642	2
59	256	860	.1925	623	1
60	.64279	.83910	1.1918	.76604	0
	cos	cot	tan	sin	′

40°

′	sin	tan	cot	cos	
0	.64279	.83910	1.1918	.76604	**60**
1	301	.83960	.1910	586	59
2	323	.84009	.1903	567	58
3	346	059	.1896	548	57
4	368	108	.1889	530	56
5	.64390	.84158	1.1882	.76511	**55**
6	412	208	.1875	492	54
7	435	258	.1868	473	53
8	457	307	.1861	455	52
9	479	357	.1854	436	51
10	.64501	.84407	1.1847	.76417	**50**
11	524	457	.1840	398	49
12	546	507	.1833	380	48
13	568	556	.1826	361	47
14	590	606	.1819	342	46
15	.64612	.84656	1.1812	.76323	**45**
16	635	706	.1806	304	44
17	657	756	.1799	286	43
18	679	806	.1792	267	42
19	701	856	.1785	248	41
20	.64723	.84906	1.1778	.76229	**40**
21	746	.84956	1771	210	39
22	768	.85006	.1764	192	38
23	790	057	.1757	173	37
24	812	107	.1750	154	36
25	.64834	.85157	1.1743	.76135	**35**
26	856	207	.1736	116	34
27	878	257	.1729	097	33
28	901	308	.1722	078	32
29	923	358	.1715	059	31
30	.64945	.85408	1.1708	.76041	**30**
31	967	458	.1702	022	29
32	.64989	509	.1695	.76003	28
33	.65011	559	.1688	.75984	27
34	033	609	.1681	965	26
35	.65055	.85660	1.1674	.75946	**25**
36	077	710	.1667	927	24
37	100	761	.1660	908	23
38	122	811	.1653	889	22
39	144	862	.1647	870	21
40	.65166	.85912	1.1640	.75851	**20**
41	188	.85963	.1633	832	19
42	210	.86014	.1626	813	18
43	232	064	.1619	794	17
44	254	115	.1612	775	16
45	.65276	.86166	1.1606	.75756	**15**
46	298	216	.1599	738	14
47	320	267	.1592	719	13
48	342	318	.1585	700	12
49	364	368	.1578	680	11
50	.65386	.86419	1.1571	.75661	**10**
51	408	470	.1565	642	9
52	430	521	.1558	623	8
53	452	572	.1551	604	7
54	474	623	.1544	585	6
55	.65496	.86674	1.1538	.75566	**5**
56	518	725	.1531	547	4
57	540	776	.1524	528	3
58	562	827	.1517	509	2
59	584	878	.1510	490	1
60	.65606	.86929	1.1504	.75471	**0**
	cos	cot	tan	sin	′

49°

41°

′	sin	tan	cot	cos	
0	.65606	.86929	1.1504	.75471	**60**
1	628	.86980	.1497	452	59
2	650	.87031	.1490	433	58
3	672	082	.1483	414	57
4	694	133	.1477	395	56
5	.65716	.87184	1.1470	.75375	**55**
6	738	236	.1463	356	54
7	759	287	.1456	337	53
8	781	338	.1450	318	52
9	803	389	.1443	299	51
10	.65825	.87441	1.1436	.75280	**50**
11	847	492	.1430	261	49
12	869	543	.1423	241	48
13	891	595	.1416	222	47
14	913	646	.1410	203	46
15	.65935	.87698	1.1403	.75184	**45**
16	956	749	.1396	165	44
17	.65978	801	.1389	146	43
18	.66000	852	.1383	126	42
19	022	904	.1376	107	41
20	.66044	.87955	1.1369	.75088	**40**
21	066	.88007	.1363	069	39
22	088	059	.1356	050	38
23	109	110	.1349	030	37
24	131	162	.1343	.75011	36
25	.66153	.88214	1.1336	.74992	**35**
26	175	265	.1329	973	34
27	197	317	.1323	953	33
28	218	369	.1316	934	32
29	240	421	.1310	915	31
30	.66262	.88473	1.1303	.74896	**30**
31	284	524	.1296	876	29
32	306	576	.1290	857	28
33	327	628	.1283	838	27
34	349	680	.1276	818	26
35	.66371	.88732	1.1270	.74799	**25**
36	393	784	.1263	780	24
37	414	836	.1257	760	23
38	436	888	.1250	741	22
39	458	940	.1243	722	21
40	.66480	.88992	1.1237	.74703	**20**
41	501	.89045	.1230	683	19
42	523	097	.1224	664	18
43	545	149	.1217	644	17
44	566	201	.1211	625	16
45	.66588	.89253	1.1204	.74606	**15**
46	610	306	.1197	586	14
47	632	358	.1191	567	13
48	653	410	.1184	548	12
49	675	463	.1178	528	11
50	.66697	.89515	1.1171	.74509	**10**
51	718	567	.1165	489	9
52	740	620	.1158	470	8
53	762	672	.1152	451	7
54	783	725	.1145	431	6
55	.66805	.89777	1.1139	.74412	**5**
56	827	830	.1132	392	4
57	848	883	.1126	373	3
58	870	935	.1119	353	2
59	891	.89988	.1113	334	1
60	.66913	.90040	1.1106	.74314	**0**
	cos	cot	tan	sin	′

48°

TABLE III

42°

′	sin	tan	cot	cos	
0	.6691$\underline{3}$	.90040	1.1106	.74314	60
1	93$\bar{5}$	093	.1100	295	59
2	956	146	.1093	276	58
3	978	199	.1087	256	57
4	.66999	251	.1080	237	56
5	.67021	.90304	1.1074	.74217	55
6	043	357	.1067	198	54
7	064	410	.1061	178	53
8	086	463	.1054	159	52
9	107	516	.1048	139	51
10	.67129	.90569	1.104$\underline{1}$	.74120	50
11	151	621	.103$\bar{5}$	100	49
12	172	674	.1028	080	48
13	194	727	.1022	061	47
14	215	781	.1016	041	46
15	.67237	.90834	1.1009	.74022	45
16	258	887	.1003	.74002	44
17	280	940	.0996	.73983	43
18	301	.90993	.0990	963	42
19	323	.91046	.0983	944	41
20	.67344	.91099	1.0977	.73924	40
21	366	153	.0971	904	39
22	387	206	.0964	88$\bar{5}$	38
23	409	259	.0958	865	37
24	430	313	.0951	846	36
25	.67452	.91366	1.0945	.73826	35
26	47$\underline{3}$	419	.0939	806	34
27	49$\bar{5}$	473	.0932	787	33
28	516	526	.0926	767	32
29	538	580	.0919	747	31
30	.67559	.91633	1.0913	.73728	30
31	580	687	.0907	708	29
32	602	740	.0900	688	28
33	62$\underline{3}$	794	.0894	669	27
34	64$\bar{5}$	847	.0888	649	26
35	.67666	.91901	1.0881	.73629	25
36	688	.9195$\bar{5}$	.0875	610	24
37	709	.92008	.0869	590	23
38	730	062	.0862	570	22
39	752	116	.0856	551	21
40	.6777$\underline{3}$	.92170	1.08$\bar{5}$0	.73531	20
41	79$\bar{5}$	224	.0843	511	19
42	816	277	.0837	491	18
43	837	331	.0831	472	17
44	859	385	.0824	452	16
45	.67880	.92439	1.0818	.73432	15
46	901	493	.0812	413	14
47	923	547	.0805	393	13
48	944	601	.0799	373	12
49	965	655	.0793	353	11
50	.67987	.92709	1.0786	.73333	10
51	.68008	763	.0780	314	9
52	029	817	.0774	294	8
53	051	872	.0768	274	7
54	072	926	.0761	254	6
55	.6809$\underline{3}$	.92980	1.0755	.73234	5
56	11$\bar{5}$	.93034	.0749	21$\bar{5}$	4
57	136	088	.0742	19$\bar{5}$	3
58	157	143	.0736	175	2
59	179	197	.0730	155	1
60	.68200	.93252	1.0724	.73135	0
	cos	cot	tan	sin	′

47°

43°

′	sin	tan	cot	cos	
0	.68200	.93252	1.0724	.73135	60
1	221	306	.0717	116	59
2	242	36$\underline{0}$	.071$\underline{1}$	096	58
3	264	41$\bar{5}$	.070$\bar{5}$	076	57
4	28$\bar{5}$	469	.0699	056	56
5	.68306	.93524	1.0692	.73036	55
6	327	578	.0686	.73016	54
7	349	633	.0680	.72996	53
8	370	688	.0674	976	52
9	391	742	.0668	957	51
10	.68412	.93797	1.0661	.72937	50
11	434	852	.0655	917	49
12	45$\bar{5}$	906	.0649	897	48
13	476	.93961	.0643	877	47
14	497	.94016	.0637	857	46
15	.68518	.94071	1.0630	.72837	45
16	539	125	.0624	817	44
17	561	180	.0618	797	43
18	582	235	.0612	777	42
19	603	290	.0606	757	41
20	.68624	.94345	1.0599	.72737	40
21	645	400	.0593	717	39
22	666	455	.0587	697	38
23	688	510	.058$\underline{1}$	677	37
24	709	565	.057$\bar{5}$	657	36
25	.68730	.94620	1.0569	.72637	35
26	751	676	.0562	617	34
27	772	731	.0556	597	33
28	793	786	.0550	577	32
29	814	841	.0544	557	31
30	.68835	.94896	1.0538	.72537	30
31	857	.94952	.0532	517	29
32	878	.95007	.0526	497	28
33	899	062	.0519	477	27
34	920	118	.0513	457	26
35	.68941	.95173	1.0507	.72437	25
36	962	229	.050$\underline{1}$	417	24
37	.68983	284	.049$\bar{5}$	397	23
38	.69004	340	.0489	377	22
39	025	395	.0483	357	21
40	.69046	.95451	1.0477	.72337	20
41	067	506	.0470	317	19
42	088	562	.0464	297	18
43	109	618	.0458	277	17
44	130	673	.0452	257	16
45	.69151	.95729	1.0446	.72236	15
46	172	78$\bar{5}$	.0440	216	14
47	193	841	.0434	196	13
48	214	897	.0428	176	12
49	235	.95952	.0422	156	11
50	.69256	.96008	1.0416	.72136	10
51	277	064	.0410	116	9
52	298	120	.0404	095	8
53	319	176	.0398	075	7
54	340	232	.0392	055	6
55	.69361	.96288	1.0385	.7203$\bar{5}$	5
56	382	344	.0379	.7201$\bar{5}$	4
57	403	400	.0373	.7199$\bar{5}$	3
58	424	457	.0367	974	2
59	44$\bar{5}$	513	.0361	954	1
60	.69466	.96569	1.0355	.71934	0
	cos	cot	tan	sin	′

46°

TABLE III **44°**

′	sin	tan	cot	cos	
0	.69466	.96569	1.0355	.71934	**60**
1	487	625	.0349	914	59
2	508	681	.0343	894	58
3	529	738	.0337	873	57
4	549	794	.0331	853	56
5	.69570	.96850	1.0325	.71833	**55**
6	591	907	.0319	813	54
7	612	.96963	.0313	792	53
8	633	.97020	.0307	772	52
9	654	076	.0301	752	51
10	.6967$\bar{5}$	.97133	1.0295	.71732	**50**
11	696	189	.0289	711	49
12	717	246	.0283	691	48
13	737	302	.0277	671	47
14	758	359	.0271	650	46
15	.69779	.97416	1.0265	.71630	**45**
16	800	472	.0259	610	44
17	821	529	.0253	590	43
18	842	586	.0247	569	42
19	862	643	.0241	549	41
20	.69883	.97700	1.0235	.71529	**40**
21	904	756	.0230	508	39
22	92$\bar{5}$	813	.0224	488	38
23	946	870	.0218	468	37
24	966	927	.0212	447	36
25	.69987	.97984	1.0206	.71427	**35**
26	.70008	.98041	.0200	407	34
27	029	098	.0194	386	33
28	049	155	.0188	366	32
29	070	213	.0182	345	31
30	.70091	.98270	1.0176	.71325	**30**
31	112	327	.0170	30$\bar{5}$	29
32	132	384	.0164	284	28
33	153	441	.0158	264	27
34	174	499	.0152	243	26
35	.7019$\bar{5}$	.98556	1.0147	.71223	**25**
36	215	613	.0141	203	24
37	236	671	.013$\bar{5}$	182	23
38	257	728	.0129	162	22
39	277	786	.0123	141	21
40	.70298	.98843	1.0117	.71121	**20**
41	319	901	.0111	100	19
42	339	.98958	.0105	080	18
43	360	.99016	.0099	059	17
44	381	073	.0094	039	16
45	.70401	.99131	1.0088	.71019	**15**
46	422	189	.0082	.70998	14
47	443	247	.0076	978	13
48	463	304	.0070	957	12
49	484	362	.0064	937	11
50	.7050$\bar{5}$	.99420	1.0058	.70916	**10**
51	525	478	.0052	896	9
52	546	536	.0047	875	8
53	567	594	.0041	85$\bar{5}$	7
54	587	652	.003$\bar{5}$	834	6
55	.70608	.99710	1.0029	.70813	**5**
56	628	768	.0023	793	4
57	649	826	.0017	772	3
58	670	884	.0012	752	2
59	690	.99942	.0006	731	1
60	.70711	1.0000	1.0000	.70711	**0**
	cos	cot	tan	sin	′

45°

TABLE IV. RADIAN MEASURE, 0° TO 180°, RADIUS = 1.

Degrees						Minutes		Seconds	
0°	0.00000 00	60°	1.04719 76	120°	2.09439 51	0′	0.00000 00	0″	0.00000 00
1	0.01745 33	61	1.06465 08	121	2.11184 84	1	0.00029 09	1	0.00000 48
2	0.03490 66	62	1.08210 41	122	2.12930 17	2	0.00058 18	2	0.00000 97
3	0.05235 99	63	1.09955 74	123	2.14675 50	3	0.00087 27	3	0.00001 45
4	0.06981 32	64	1.11701 07	124	2.16420 83	4	0.00116 36	4	0.00001 94
5	0.08726 65	**65**	1.13446 40	**125**	2.18166 16	**5**	0.00145 44	**5**	0.00002 42
6	0.10471 98	66	1.15191 73	126	2.19911 49	6	0.00174 53	6	0.00002 91
7	0.12217 30	67	1.16937 06	127	2.21656 82	7	0.00203 62	7	0.00003 39
8	0.13962 63	68	1.18682 39	128	2.23402 14	8	0.00232 71	8	0.00003 88
9	0.15707 96	69	1.20427 72	129	2.25147 47	9	0.00261 80	9	0.00004 36
10	0.17453 29	**70**	1.22173 05	**130**	2.26892 80	**10**	0.00290 89	**10**	0.00004 85
11	0.19198 62	71	1.23918 38	131	2.28638 13	11	0.00319 98	11	0.00005 33
12	0.20943 95	72	1.25663 71	132	2.30383 46	12	0.00349 07	12	0.00005 82
13	0.22689 28	73	1.27409 04	133	2.32128 79	13	0.00378 15	13	0.00006 30
14	0.24434 61	74	1.29154 36	134	2.33874 12	14	0.00407 24	14	0.00006 79
15	0.26179 94	**75**	1.30899 69	**135**	2.35619 45	**15**	0.00436 33	**15**	0.00007 27
16	0.27925 27	76	1.32645 02	136	2.37364 78	16	0.00465 42	16	0.00007 76
17	0.29670 60	77	1.34390 35	137	2.39110 11	17	0.00494 51	17	0.00008 24
18	0.31415 93	78	1.36135 68	138	2.40855 44	18	0.00523 60	18	0.00008 73
19	0.33161 26	79	1.37881 01	139	2.42600 77	19	0.00552 69	19	0.00009 21
20	0.34906 59	**80**	1.39626 34	**140**	2.44346 10	**20**	0.00581 78	**20**	0.00009 70
21	0.36651 91	81	1.41371 67	141	2.46091 42	21	0.00610 87	21	0.00010 18
22	0.38397 24	82	1.43117 00	142	2.47836 75	22	0.00639 95	22	0.00010 67
23	0.40142 57	83	1.44862 33	143	2.49582 08	23	0.00669 04	23	0.00011 15
24	0.41887 90	84	1.46607 66	144	2.51327 41	24	0.00698 13	24	0.00011 64
25	0.43633 23	85	1.48352 99	**145**	2.53072 74	**25**	0.00727 22	**25**	0.00012 12
26	0.45378 56	86	1.50098 32	146	2.54818 07	26	0.00756 31	26	0.00012 61
27	0.47123 89	87	1.51843 64	147	2.56563 40	27	0.00785 40	27	0.00013 09
28	0.48869 22	88	1.53588 97	148	2.58308 73	28	0.00814 49	28	0.00013 57
29	0.50614 55	89	1.55334 30	149	2.60054 06	29	0.00843 58	29	0.00014 06
30	0.52359 88	**90**	1.57079 63	**150**	2.61799 39	**30**	0.00872 66	**30**	0.00014 54
31	0.54105 21	91	1.58824 96	151	2.63544 72	31	0.00901 75	31	0.00015 03
32	0.55850 54	92	1.60570 29	152	2.65290 05	32	0.00930 84	32	0.00015 51
33	0.57595 87	93	1.62315 62	153	2.67035 38	33	0.00959 93	33	0.00016 00
34	0.59341 19	94	1.64060 95	154	2.68780 70	34	0.00989 02	34	0.00016 48
35	0.61086 52	**95**	1.65806 28	**155**	2.70526 03	**35**	0.01018 11	**35**	0.00016 97
36	0.62831 85	96	1.67551 61	156	2.72271 36	36	0.01047 20	36	0.00017 45
37	0.64577 18	97	1.69296 94	157	2.74016 69	37	0.01076 29	37	0.00017 94
38	0.66322 51	98	1.71042 27	158	2.75762 02	38	0.01105 38	38	0.00018 42
39	0.68067 84	99	1.72787 60	159	2.77507 35	39	0.01134 46	39	0.00018 91
40	0.69813 17	**100**	1.74532 93	**160**	2.79252 68	**40**	0.01163 55	**40**	0.00019 39
41	0.71558 50	101	1.76278 25	161	2.80998 01	41	0.01192 64	41	0.00019 88
42	0.73303 83	102	1.78023 58	162	2.82743 34	42	0.01221 73	42	0.00020 36
43	0.75049 16	103	1.79768 91	163	2.84488 67	43	0.01250 82	43	0.00020 85
44	0.76794 49	104	1.81514 24	164	2.86234 00	44	0.01279 91	44	0.00021 33
45	0.78539 82	**105**	1.83259 57	**165**	2.87979 33	**45**	0.01309 00	**45**	0.00021 82
46	0.80285 15	106	1.85004 90	166	2.89724 66	46	0.01338 09	46	0.00022 30
47	0.82030 47	107	1.86750 23	167	2.91469 99	47	0.01367 17	47	0.00022 79
48	0.83775 80	108	1.88495 56	168	2.93215 31	48	0.01396 26	48	0.00023 27
49	0.85521 13	109	1.90240 89	169	2.94960 64	49	0.01425 35	49	0.00023 76
50	0.87266 46	**110**	1.91986 22	**170**	2.96705 97	**50**	0.01454 44	**50**	0.00024 24
51	0.89011 79	111	1.93731 55	171	2.98451 30	51	0.01483 53	51	0.00024 73
52	0.90757 12	112	1.95476 88	172	3.00196 63	52	0.01512 62	52	0.00025 21
53	0.92502 45	113	1.97222 21	173	3.01941 96	53	0.01541 71	53	0.00025 70
54	0.94247 78	114	1.98967 53	174	3.03687 29	54	0.01570 80	54	0.00026 18
55	0.95993 11	**115**	2.00712 86	**175**	3.05432 62	**55**	0.01599 89	**55**	0.00026 66
56	0.97738 44	116	2.02458 19	176	3.07177 95	56	0.01628 97	56	0.00027 15
57	0.99483 77	117	2.04203 52	177	3.08923 28	57	0.01658 06	57	0.00027 63
58	1.01229 10	118	2.05948 85	178	3.10668 61	58	0.01687 15	58	0.00028 12
59	1.02974 43	119	2.07694 18	179	3.12413 94	59	0.01716 24	59	0.00028 60
60	1.04719 76	**120**	2.09439 51	**180**	3.14159 27	**60**	0.01745 33	**60**	0.00029 09
Degrees						Minutes		Seconds	

TABLE V. HAVERSINES

DEGREES	0′		10′		20′		30′		40′		50′	
	l hav	n hav	*l hav*	n hav	*l hav*	n hav	*l hav*	n hav	*l hav*	n hav	*l hav*	n hav
	Obtain characteristic of l hav by inspection of n hav											
0	—	**0000**	3254	**0000**	9275	**0000**	2796	**0000**	5295	**0000**	7233	**0001**
1	8817	**0001**	0156	**0001**	1316	**0001**	2339	**0002**	3254	**0002**	4081	**0003**
2	4837	**0003**	5532	**0004**	6176	**0004**	6775	**0005**	7336	**0005**	7862	**0006**
3	8358	**0007**	8828	**0008**	9273	**0008**	9697	**0009**	0101	**0010**	0487	**0011**
4	0856	**0012**	1211	**0013**	1551	**0014**	1879	**0015**	2195	**0017**	2499	**0018**
5	2794	**0019**	3078	**0020**	3354	**0022**	3621	**0023**	3880	**0024**	4132	**0026**
6	4376	**0027**	4614	**0029**	4845	**0031**	5071	**0032**	5290	**0034**	5504	**0036**
7	5714	**0037**	5918	**0039**	6117	**0041**	6312	**0043**	6503	**0045**	6689	**0047**
8	6872	**0049**	7051	**0051**	7226	**0053**	7397	**0055**	7566	**0057**	7731	**0059**
9	7893	**0062**	8052	**0064**	8208	**0066**	8361	**0069**	8512	**0071**	8660	**0073**
10	8806	**0076**	8949	**0079**	9090	**0081**	9229	**0084**	9365	**0086**	9499	**0089**
11	9631	**0092**	9762	**0095**	9890	**0097**	0016	**0100**	0141	**0103**	0264	**0106**
12	0385	**0109**	0504	**0112**	0622	**0115**	0738	**0119**	0852	**0122**	0966	**0125**
13	1077	**0128**	1187	**0131**	1296	**0135**	1404	**0138**	1510	**0142**	1614	**0145**
14	1718	**0149**	1820	**0152**	1921	**0156**	2021	**0159**	2120	**0163**	2217	**0167**
15	2314	**0170**	2409	**0174**	2504	**0178**	2597	**0182**	2689	**0186**	2781	**0190**
16	2871	**0194**	2961	**0198**	3049	**0202**	3137	**0206**	3223	**0210**	3309	**0214**
17	3394	**0218**	3478	**0223**	3561	**0227**	3644	**0231**	3726	**0236**	3807	**0240**
18	3887	**0245**	3966	**0249**	4045	**0254**	4123	**0258**	4200	**0263**	4276	**0268**
19	4352	**0272**	4427	**0277**	4502	**0282**	4576	**0287**	4649	**0292**	4721	**0297**
20	4793	**0302**	4865	**0307**	4935	**0312**	5006	**0317**	5075	**0322**	5144	**0327**
21	5213	**0332**	5281	**0337**	5348	**0343**	5415	**0348**	5481	**0353**	5547	**0359**
22	5612	**0364**	5677	**0370**	5741	**0375**	5805	**0381**	5868	**0386**	5931	**0392**
23	5993	**0397**	6055	**0403**	6116	**0409**	6177	**0415**	6238	**0421**	6298	**0426**
24	6358	**0432**	6417	**0438**	6476	**0444**	6534	**0450**	6592	**0456**	6650	**0462**
25	6707	**0468**	6764	**0475**	6820	**0481**	6876	**0487**	6932	**0493**	6987	**0500**
26	7042	**0506**	7096	**0512**	7150	**0519**	7204	**0525**	7258	**0532**	7311	**0538**
27	7364	**0545**	7416	**0552**	7468	**0558**	7520	**0565**	7572	**0572**	7623	**0578**
28	7674	**0585**	7724	**0592**	7774	**0599**	7824	**0606**	7874	**0613**	7923	**0620**
29	7972	**0627**	8021	**0634**	8069	**0641**	8117	**0648**	8165	**0655**	8213	**0663**
30	8260	**0670**	8307	**0677**	8354	**0684**	8400	**0692**	8446	**0699**	8492	**0707**
31	8538	**0714**	8583	**0722**	8629	**0729**	8673	**0737**	8718	**0744**	8763	**0752**
32	8807	**0760**	8851	**0767**	8894	**0775**	8938	**0783**	8981	**0791**	9024	**0799**
33	9067	**0807**	9109	**0815**	9152	**0823**	9194	**0831**	9236	**0839**	9277	**0847**
34	9319	**0855**	9360	**0863**	9401	**0871**	9442	**0879**	9482	**0888**	9523	**0896**
35	9563	**0904**	9603	**0913**	9643	**0921**	9682	**0929**	9721	**0938**	9761	**0946**
36	9800	**0955**	9838	**0963**	9877	**0972**	9915	**0981**	9954	**0989**	9992	**0998**
37	0030	**1007**	0067	**1016**	0105	**1024**	0142	**1033**	0179	**1042**	0216	**1051**
38	0253	**1060**	0289	**1069**	0326	**1078**	0362	**1087**	0398	**1096**	0434	**1105**
39	0470	**1114**	0505	**1123**	0541	**1133**	0576	**1142**	0611	**1151**	0646	**1160**
40	0681	**1170**	0716	**1179**	0750	**1189**	0784	**1198**	0819	**1207**	0853	**1217**
41	0887	**1226**	0920	**1236**	0954	**1246**	0987	**1255**	1020	**1265**	1054	**1275**
42	1087	**1284**	1119	**1294**	1152	**1304**	1185	**1314**	1217	**1323**	1249	**1333**
43	1282	**1343**	1314	**1353**	1345	**1363**	1377	**1373**	1409	**1383**	1440	**1393**
44	1472	**1403**	1503	**1413**	1534	**1424**	1565	**1434**	1596	**1444**	1626	**1454**
45	1657	**1464**	1687	**1475**	1718	**1485**	1748	**1495**	1778	**1506**	1808	**1516**
46	1838	**1527**	1867	**1537**	1897	**1548**	1926	**1558**	1956	**1569**	1985	**1579**
47	2014	**1590**	2043	**1601**	2072	**1611**	2101	**1622**	2129	**1633**	2158	**1644**
48	2186	**1654**	2215	**1665**	2243	**1676**	2271	**1687**	2299	**1698**	2327	**1709**
49	2355	**1720**	2382	**1731**	2410	**1742**	2437	**1753**	2465	**1764**	2492	**1775**
50	2519	**1786**	2546	**1797**	2573	**1808**	2600	**1820**	2627	**1831**	2653	**1842**
51	2680	**1853**	2706	**1865**	2732	**1876**	2759	**1887**	2785	**1899**	2811	**1910**
52	2837	**1922**	2863	**1933**	2888	**1945**	2914	**1956**	2940	**1968**	2965	**1979**
53	2991	**1991**	3016	**2003**	3041	**2014**	3066	**2026**	3091	**2038**	3116	**2049**
54	3141	**2061**	3166	**2073**	3190	**2085**	3215	**2096**	3239	**2108**	3264	**2120**
55	3288	**2132**	3312	**2144**	3336	**2156**	3361	**2168**	3384	**2180**	3408	**2192**
56	3432	**2204**	3456	**2216**	3480	**2228**	3503	**2240**	3527	**2252**	3550	**2265**
57	3573	**2277**	3596	**2289**	3620	**2301**	3643	**2314**	3666	**2326**	3689	**2338**
58	3711	**2350**	3734	**2363**	3757	**2375**	3779	**2388**	3802	**2400**	3824	**2412**
59	3847	**2425**	3869	**2437**	3891	**2450**	3913	**2462**	3935	**2475**	3957	**2487**

DEGREES	0′		10′		20′		30′		40′		50′	
	l hav 9.	n hav	*l hav* 9.	n hav	*l hav* 9.	n hav	*l hav* 9.	n hav	*l hav* 9.	n hav	*l hav* 9.	n hav
60	3979	2500	4001	2513	4023	2525	4045	2538	4066	2551	4088	2563
61	4109	2576	4131	2589	4152	2601	4173	2614	4195	2627	4216	2640
62	4237	2653	4258	2665	4279	2678	4300	2691	4320	2704	4341	2717
63	4362	2730	4382	2743	4403	2756	4423	2769	4444	2782	4464	2795
64	4484	2808	4504	2821	4524	2834	4545	2847	4565	2861	4584	2874
65	4604	2887	4624	2900	4644	2913	4664	2927	4683	2940	4703	2953
66	4722	2966	4742	2980	4761	2993	4780	3006	4799	3020	4819	3033
67	4838	3046	4857	3060	4876	3073	4895	3087	4914	3100	4932	3113
68	4951	3127	4970	3140	4989	3154	5007	3167	5026	3181	5044	3195
69	5063	3208	5081	3222	5099	3235	5117	3249	5136	3263	5154	3276
70	5172	3290	5190	3304	5208	3317	5226	3331	5244	3345	5261	3358
71	5279	3372	5297	3386	5314	3400	5332	3413	5349	3427	5367	3441
72	5384	3455	5402	3469	5419	3483	5436	3496	5454	3510	5471	3524
73	5488	3538	5505	3552	5522	3566	5539	3580	5556	3594	5572	3608
74	5589	3622	5606	3636	5623	3650	5639	3664	5656	3678	5672	3692
75	5689	3706	5705	3720	5722	3734	5738	3748	5754	3762	5771	3776
76	5787	3790	5803	3805	5819	3819	5835	3833	5851	3847	5867	3861
77	5883	3875	5899	3889	5915	3904	5930	3918	5946	3932	5962	3946
78	5977	3960	5993	3975	6009	3989	6024	4003	6039	4017	6055	4032
79	6070	4046	6086	4060	6101	4075	6116	4089	6131	4103	6146	4117
80	6161	4132	6176	4146	6191	4160	6206	4175	6221	4189	6236	4203
81	6251	4218	6266	4232	6280	4247	6295	4261	6310	4275	6324	4290
82	6339	4304	6353	4319	6368	4333	6382	4347	6397	4362	6411	4376
83	6425	4391	6440	4405	6454	4420	6468	4434	6482	4448	6496	4463
84	6510	4477	6524	4492	6538	4506	6552	4521	6566	4535	6580	4550
85	6594	4564	6607	4579	6621	4593	6635	4608	6648	4622	6662	4637
86	6676	4651	6689	4666	6703	4680	6716	4695	6730	4709	6743	4724
87	6756	4738	6770	4753	6783	4767	6796	4782	6809	4796	6822	4811
88	6835	4826	6848	4840	6862	4855	6875	4869	6887	4884	6900	4898
89	6913	4913	6926	4927	6939	4942	6952	4956	6964	4971	6977	4985
90	6990	5000	7002	5015	7015	5029	7027	5044	7040	5058	7052	5073
91	7065	5087	7077	5102	7090	5116	7102	5131	7114	5145	7126	5160
92	7139	5174	7151	5189	7163	5204	7175	5218	7187	5233	7199	5247
93	7211	5262	7223	5276	7235	5291	7247	5305	7259	5320	7271	5334
94	7283	5349	7294	5363	7306	5378	7318	5392	7329	5407	7341	5421
95	7353	5436	7364	5450	7376	5465	7387	5479	7399	5494	7410	5508
96	7421	5523	7433	5537	7444	5552	7455	5566	7467	5580	7478	5595
97	7489	5609	7500	5624	7511	5638	7523	5653	7534	5667	7545	5681
98	7556	5696	7567	5710	7577	5725	7588	5739	7599	5753	7610	5768
99	7621	5782	7632	5797	7642	5811	7653	5825	7664	5840	7674	5854
100	7685	5868	7696	5883	7706	5897	7717	5911	7727	5925	7738	5940
101	7748	5954	7759	5968	7769	5983	7779	5997	7790	6011	7800	6025
102	7810	6040	7820	6054	7830	6068	7841	6082	7851	6096	7861	6111
103	7871	6125	7881	6139	7891	6153	7901	6167	7911	6181	7921	6195
104	7931	6210	7940	6224	7950	6238	7960	6252	7970	6266	7980	6280
105	7989	6294	7999	6308	8009	6322	8018	6336	8028	6350	8037	6364
106	8047	6378	8056	6392	8066	6406	8075	6420	8085	6434	8094	6448
107	8104	6462	8113	6476	8122	6490	8131	6504	8141	6517	8150	6531
108	8159	6545	8168	6559	8177	6573	8187	6587	8196	6600	8205	6614
109	8214	6628	8223	6642	8232	6655	8241	6669	8250	6683	8258	6696
110	8267	6710	8276	6724	8285	6737	8294	6751	8302	6765	8311	6778
111	8320	6792	8329	6805	8337	6819	8346	6833	8354	6846	8363	6860
112	8371	6873	8380	6887	8388	6900	8397	6913	8405	6927	8414	6940
113	8422	6954	8430	6967	8439	6980	8447	6994	8455	7007	8464	7020
114	8472	7034	8480	7047	8488	7060	8496	7073	8504	7087	8513	7100
115	8521	7113	8529	7126	8537	7139	8545	7153	8553	7166	8561	7179
116	8568	7192	8576	7205	8584	7218	8592	7231	8600	7244	8608	7257
117	8615	7270	8623	7283	8631	7296	8638	7309	8646	7322	8654	7335
118	8661	7347	8669	7360	8676	7373	8684	7386	8691	7399	8699	7411
119	8706	7424	8714	7437	8721	7449	8729	7462	8736	7475	8743	7487

DEGREES	0′		10′		20′		30′		40′		50′	
	l hav 9.	n hav	*l hav* 9.	n hav	*l hav* 9.	n hav	*l hav* 9.	n hav	*l hav* 9.	n hav	*l hav* 9.	n hav
120	8751	**7500**	8758	**7513**	8765	**7525**	8772	**7538**	8780	**7550**	8787	**7563**
121	8794	**7575**	8801	**7588**	8808	**7600**	8815	**7612**	8822	**7625**	8829	**7637**
122	8836	**7650**	8843	**7662**	8850	**7674**	8857	**7686**	8864	**7699**	8871	**7711**
123	8878	**7723**	8885	**7735**	8892	**7748**	8898	**7760**	8905	**7772**	8912	**7784**
124	8919	**7796**	8925	**7808**	8932	**7820**	8939	**7832**	8945	**7844**	8952	**7856**
125	8959	**7868**	8965	**7880**	8972	**7892**	8978	**7904**	8985	**7915**	8991	**7927**
126	8998	**7939**	9004	**7951**	9010	**7962**	9017	**7974**	9023	**7986**	9030	**7997**
127	9036	**8009**	9042	**8021**	9048	**8032**	9055	**8044**	9061	**8055**	9067	**8067**
128	9073	**8078**	9079	**8090**	9085	**8101**	9092	**8113**	9098	**8124**	9104	**8135**
129	9110	**8147**	9116	**8158**	9122	**8169**	9128	**8180**	9134	**8192**	9140	**8203**
130	9146	**8214**	9151	**8225**	9157	**8236**	9163	**8247**	9169	**8258**	9175	**8269**
131	9180	**8280**	9186	**8291**	9192	**8302**	9198	**8313**	9203	**8324**	9209	**8335**
132	9215	**8346**	9220	**8356**	9226	**8367**	9231	**8378**	9237	**8389**	9242	**8399**
133	9248	**8410**	9253	**8421**	9259	**8431**	9264	**8442**	9270	**8452**	9275	**8463**
134	9281	**8473**	9286	**8484**	9291	**8494**	9297	**8505**	9302	**8515**	9307	**8525**
135	9312	**8536**	9318	**8546**	9323	**8556**	9328	**8566**	9333	**8576**	9338	**8587**
136	9343	**8597**	9348	**8607**	9353	**8617**	9359	**8627**	9364	**8637**	9369	**8647**
137	9374	**8657**	9379	**8667**	9383	**8677**	9388	**8686**	9393	**8696**	9398	**8706**
138	9403	**8716**	9408	**8725**	9413	**8735**	9417	**8745**	9422	**8754**	9427	**8764**
139	9432	**8774**	9436	**8783**	9441	**8793**	9446	**8802**	9450	**8811**	9455	**8821**
140	9460	**8830**	9464	**8840**	9469	**8849**	9473	**8858**	9478	**8867**	9482	**8877**
141	9487	**8886**	9491	**8895**	9496	**8904**	9500	**8913**	9505	**8922**	9509	**8931**
142	9513	**8940**	9518	**8949**	9522	**8958**	9526	**8967**	9531	**8976**	9535	**8984**
143	9539	**8993**	9543	**9002**	9548	**9011**	9552	**9019**	9556	**9028**	9560	**9037**
144	9564	**9045**	9568	**9054**	9572	**9062**	9576	**9071**	9580	**9079**	9584	**9087**
145	9588	**9096**	9592	**9104**	9596	**9112**	9600	**9121**	9604	**9129**	9608	**9137**
146	9612	**9145**	9616	**9153**	9620	**9161**	9623	**9169**	9627	**9177**	9631	**9185**
147	9635	**9193**	9638	**9201**	9642	**9209**	9646	**9217**	9650	**9225**	9653	**9233**
148	9657	**9240**	9660	**9248**	9664	**9256**	9668	**9263**	9671	**9271**	9675	**9278**
149	9678	**9286**	9682	**9293**	9685	**9301**	9689	**9308**	9692	**9316**	9695	**9323**
150	9699	**9330**	9702	**9337**	9706	**9345**	9709	**9352**	9712	**9359**	9716	**9366**
151	9719	**9373**	9722	**9380**	9725	**9387**	9729	**9394**	9732	**9401**	9735	**9408**
152	9738	**9415**	9741	**9422**	9744	**9428**	9747	**9435**	9751	**9442**	9754	**9448**
153	9757	**9455**	9760	**9462**	9763	**9468**	9766	**9475**	9769	**9481**	9772	**9488**
154	9774	**9494**	9777	**9500**	9780	**9507**	9783	**9513**	9786	**9519**	9789	**9525**
155	9792	**9532**	9794	**9538**	9797	**9544**	9800	**9550**	9803	**9556**	9805	**9562**
156	9808	**9568**	9811	**9574**	9813	**9579**	9816	**9585**	9819	**9591**	9821	**9597**
157	9824	**9603**	9826	**9608**	9829	**9614**	9831	**9619**	9834	**9625**	9836	**9630**
158	9839	**9636**	9841	**9641**	9844	**9647**	9846	**9652**	9849	**9657**	9851	**9663**
159	9853	**9668**	9856	**9673**	9858	**9678**	9860	**9683**	9863	**9688**	9865	**9693**
160	9867	**9698**	9869	**9703**	9871	**9708**	9874	**9713**	9876	**9718**	9878	**9723**
161	9880	**9728**	9882	**9732**	9884	**9737**	9886	**9742**	9888	**9746**	9890	**9751**
162	9892	**9755**	9894	**9760**	9896	**9764**	9898	**9769**	9900	**9773**	9902	**9777**
163	9904	**9782**	9906	**9786**	9908	**9790**	9910	**9794**	9911	**9798**	9913	**9802**
164	9915	**9806**	9917	**9810**	9919	**9814**	9920	**9818**	9922	**9822**	9924	**9826**
165	9925	**9830**	9927	**9833**	9929	**9837**	9930	**9841**	9932	**9844**	9933	**9848**
166	9935	**9851**	9937	**9855**	9938	**9858**	9940	**9862**	9941	**9865**	9943	**9869**
167	9944	**9872**	9945	**9875**	9947	**9878**	9948	**9881**	9950	**9885**	9951	**9888**
168	9952	**9891**	9954	**9894**	9955	**9897**	9956	**9900**	9957	**9903**	9959	**9905**
169	9960	**9908**	9961	**9911**	9962	**9914**	9963	**9916**	9965	**9919**	9966	**9921**
170	9967	**9924**	9968	**9927**	9969	**9929**	9970	**9931**	9971	**9934**	9972	**9936**
171	9973	**9938**	9974	**9941**	9975	**9943**	9976	**9945**	9977	**9947**	9978	**9949**
172	9979	**9951**	9980	**9953**	9981	**9955**	9981	**9957**	9982	**9959**	9983	**9961**
173	9984	**9963**	9985	**9964**	9985	**9966**	9986	**9968**	9987	**9969**	9987	**9971**
174	9988	**9973**	9989	**9974**	9989	**9976**	9990	**9977**	9991	**9978**	9991	**9980**
175	9992	**9981**	9992	**9982**	9993	**9983**	9993	**9985**	9994	**9986**	9994	**9987**
176	9995	**9988**	9995	**9989**	9996	**9990**	9996	**9991**	9996	**9992**	9997	**9992**
177	9997	**9993**	9997	**9994**	9998	**9995**	9998	**9995**	9998	**9996**	9998	**9996**
178	9999	**9997**	9999	**9997**	9999	**9998**	9999	**9998**	9999	**9999**	9999	**9999**
179	9999	**9999**	9999	**9999**	0000	**0000**	0000	**0000**	0000	**0000**	0000	**0000**
180	0000	**0000**										

Catalog

If you are interested in a list of fine Paperback books, covering a wide range of subjects and interests, send your name and address, requesting your free catalog, to: